Frederick W Parham

Thoracic Resection for tTumors Growing

From the Bony Wall of the Chest

Frederick W Parham

Thoracic Resection for tTumors Growing
From the Bony Wall of the Chest

ISBN/EAN: 9783337021740

Printed in Europe, USA, Canada, Australia, Japan

Cover: Foto ©berggeist007 / pixelio.de

More available books at **www.hansebooks.com**

THORACIC RESECTION FOR TUMORS GROWING FROM THE BONY WALL OF THE CHEST.

BY

F. W. PARHAM, M.D.,

PROFESSOR OF GENERAL CLINICAL AND OPERATIVE SURGERY,
NEW ORLEANS POLYCLINIC.

Read in Abstract before the
Southern Surgical and Gynecological Association,
Memphis, Nov., 1898.

NEW ORLEANS.
1899.

CONTENTS.

THORACIC RESECTION FOR TUMORS GROW-
ING FROM THE BONY WALL OF
THE CHEST.

"All experience is an arch wherethro'
Gleams that untravelled world, whose margin fades
Forever and forever when I move."—ULYSSES.

IN the spring of 1897 I operated in the Charity Hospital upon a young colored woman for a sarcoma of the thoracic wall requiring the resection of several inches of the third, fourth, and fifth ribs. It was impossible to separate the tumor from the pleura without tearing the latter; a rent of five inches was made, which permitted easily the thrusting of the whole hand into the pleural cavity. Suddenly was presented to our anxious view one of the most startling clinical pictures that the surgeon can ever be called upon to witness. At such a sight the stoutest heart will quaver. No wonder the older surgeons discountenanced such operations, and in the times when there was neither anæsthetic nor asepsis the surgeon might well be excused for indorsing the condemnation uttered by Dieffenbach against Milton Antony for his temerity in venturing to remove a tumor of the chest-wall with a large piece of the lung. This Dieffenbach pronounced the greatest error in this field of surgery, and he added that "the like operations go too much counter to all physiology and give to the beautiful saving art the appearance of cruel annihilation-mechanics."

So sudden in my case was the pneumothorax and so striking

were the manifestations of profound shock, threatening almost
instant dissolution before our eyes, that I resolved to acquaint
myself more thoroughly with the dangers of thoracic surgery
and to study the resources of our art for successfully dealing
with such emergencies as are born with the sudden, accidental,
or intentional, opening of the pleural cavity.

Soon after that my friend, Dr. Matas, chairman of the
Surgical Section of the Louisiana State Medical Society,
selected for his address the subject of thoracic surgery, and
I agreed to open the discussion. Dr. Matas at once took up,
with his usual vigor and thoroughness, the investigation of
the whole field of chest surgery. His work, soon to be published, presents in a most interesting and thorough manner
a review of the astounding progress made in the past twenty
years in this important branch, and clearly indicates its glow-
ing possibilities in the coming decades.

The result of his labors lies before me. I shall strive in
this paper not to trench too much upon his ground, but shall
rather try to provide the complement to his work, in so far
as it relates to but one phase of thoracic surgery, as indicated
in the title of this paper.

Since the meeting of the State Medical Society, in May,
1898, when the discussion on chest-surgery took place, I have
had another case, similar to the one above referred to; but,
although both happily recovered, the almost insignificant disturbances during and after the operation in the last case were
in such striking contrast with the stormy manifestations of my
first operative attack that I cannot forbear here to express
my deep gratitude to Fell and O'Dwyer, whose labors in the
fields of intubation and forced respiration enabled me in the
second case to overcome the dangers of the first. I would
not willingly discredit in any degree the original investigations of Tuffier and Hallion and of Quénu and Longuet,
nor would I speak disparagingly of Doyen's mechanical inge-
nuity as displayed in his independently devised apparatus for
artificial respiration; but it is only just to say that credit
belongs to Dr. Fell, of Buffalo, for giving to surgery an

apparatus embodying in its practical evolution the principles so ardently urged by Fell and O'Dwyer. As far as I can learn, I am the first to demonstrate the value of this admirable apparatus (used in my second case) in maintaining the respiration during operations of this kind.

It is my purpose in this paper to present all the cases of resection for malignant tumor of the thoracic skeleton that I have been able to find in medical literature.

My work has been much facilitated by the chapter on this subject in Paget's entertaining work on the *Surgery of the Chest*, 1896 ; by Quénu and Longuet's article " Des Tumeurs du Squelette Thoracique," in the *Revue de Chirurgie*, May 10, 1898 ; by the paper of Gerulanos on " Operative Pneumothorax," in the *Deutsch. Zeitschr. f. Chir.*, for October, 1898. To these I am especially indebted, but in most cases I have consulted the original bibliographic references, and have thus been enabled to correct some errors in reports of cases, to throw out some erroneously included, and to add others not found in any of the collections of cases accessible to me.

I regret very much that in the preparation of this paper I was unable to find in the Surgeon General's Library at Washington, Campe's *Göttingen-Dissertation*, 1894, " Ueber Tumoren der Knochernen Thoraxwand," in which he has collected fifty-seven cases, of which forty were costal and seventeen sternal. I particularly regret my failure to see his list, because I should have been able to get fuller accounts of some, especially Kœnig's interesting cases. I have, however, succeeded in obtaining many of his data from Paget's work, where Campe's thesis is properly credited.

Other references not found in connection with cases reported will be found at the end of this paper.

Tumors of the ribs and sternum may be conveniently divided into :

Primary, including fibroma, chondroma, sarcoma, and their combinations.

Secondary, including sarcomas and carcinomas.

Quénu and Longuet have tabulated their cases in two classes

—the extrapleural or non-penetrating, and the intrapleural or penetrating. They have further arranged them under fibromata, enchondromata, and sarcomata, excluding altogether the carcinomata.

I have adopted the division into extrapleural and intrapleural operations, but have not attempted separate grouping of the histological forms, and have included carcinomas, believing these should enter into a determination of the operative results, and should no more be excluded from such a list than many of the malignant sarcomas.

These cases are arranged by years, so as to exhibit the movement in surgical progress to better advantage.

The operative attack must be so much modified, according as to whether the pleura is involved or not, that I shall consider the cases in two distinct series.

Series I. Cases in which the pleura was not opened.

Series II. Cases in which the pleura was opened either deliberately or accidentally.

Series I. Cases in which the Pleural Cavity was not Opened.

Case I.

Case of Osias Aimar, 1778, osteosarcoma. This case is erroneously made to appear as two cases in Quénu and Longuet's table. No case can be properly credited to Lazarius Riverius, as none is to be found in his works except that communicated to him by Osias Aimar. It appears as the second case in observation No. 491 in the list of 513 observations or histories in the *Four Books of Lazarius Riverius*, translated by Nicholas Culpepper, London, 1678, p. 315. The case is as follows :

"M. de Beflin, captain of a band of soldiers, had for a long time a scirrhous or hard, senseless tumor in his left side upon the true ribs, viz., the fifth, sixth, and seventh. A certain surgeon opened the tumor with an actual cautery, out of which there came very little matter, and there remained very grievous pains in the part; which compelled him, living in the country, to come to Grenoble and implore my help. I found the ulcer as big as the palm of one's hand and the ribs underneath infected more than half their length with rottenness. Premising, therefore, universal medicaments, I cut off the extremities of the ribs the quantity of four fingers' breadth, applying afterward an actual cautery to their extremities. Finally with catagmatical powders I procured a separation of the burned parts, and with detergent and incarnating medicines I did heal up the ulcer and bring it to a scar."

This case is probably correctly classified by Quénu and Longuet as one of osteosarcoma, but the description does not justify one in saying that it may not have been tuberculous disease of the ribs or carcinoma. The case is, however, retained.

CASE II.

Case of Cittadini, 1820, osteosarcoma (*Annali Universali de medicina,* Milano, 1826, xxxvii., 404).

F. L., male (age not stated), farm laborer, had suffered for many years with a tumor, which was now fungous in character and situated in the cartilages of the sixth and seventh true ribs and in the first false rib an inch from xiphoid appendix. Its base did not exceed two inches in circumference.

The operation was done in February, 1820, by Cittadini.

A piece of skin around the fungus was removed, three inches in diameter, and the muscular fibres, belonging partly to the rectus, partly to the great abdominal oblique, loosened by blunt dissection; and the fungus mass was removed together with the cartilages and rib involved. Hemorrhage was arrested by fire. It would appear from the account that a fungous growth was left behind, but it is nevertheless stated that, when the man died two years afterward, the fungous growth had not reappeared.

The healing was slow, but was finally completed after four months.

CASE III.

Warren's case, 1837, osteochondroma (*Boston Medical and Surgical Journal*, 1837, vol. xvi. p. 128).

The patient was formerly a shoemaker, aged forty years, but at present is a trader. He came to Boston in the spring of 1836 to consult Dr. Warren for a hard tumor on the cartilages of the left ribs. His case was this : Two years before, while holding a shoe, on which he was at work, against the breast, the shoe slipped under the pressure and struck him on the cartilage of the sixth rib. The following day he perceived a slight swelling on the spot. This gradually increased, and was about three inches in length when he was first examined. An operation was advised, to which he assented; but having some business to transact he returned home, and did not appear again until about the commencement of the present month, when he entered the hospital for the purpose of submitting himself to an operation. The size of the tumor during this period had greatly increased. Beginning at the upper edge of the cartilage of the fifth rib, it extended to the lower edge of that of the eighth, being about five inches in length from above downward, and four in a transverse direction, from the median line to the left. In color the skin was not changed, excepting that it exhibited numerous enlarged veins. In consistence it had the firmness of a periosteal tumor —that is, something less than a bony hardness. A degree of sensibility existed on the edge of the cartilage of the fifth rib and at some other points. It was slightly movable in a lateral direction, but not in the vertical, and its movement did not appear to affect the ribs. There was a sensible pulsation in it without vibration. The patient wished to know whether an operation for his relief could be safely performed. The skin over the tumor was adherent.

The first question was, What is the nature of this tumor? To the eye it had the appearance of an osteosarcoma ; to the touch it wanted the osseous plates of that disease. Was it a disease of the perichondrium ? Might it not be a projection caused by internal aneurism? Did it lie on the outside of the ribs, or might it not extend inward as well as outward? There was an obvious dip of the tumor below the edges of the cartilages into the epigastrium.

Dr. Warren seemed to think that it was either a scirrhous or perichondrial disease—lying under the externus oblique and rectus muscles, and over the cartilages of the ribs, whence it descended into the epigastric region and came in contact with the external face of the internal oblique. As, however, it might extend through the thickness of the cartilages, it would be necessary to consider the possibility of taking out these cartilages. It appeared to be practicable, after cutting through the externus oblique and transversalis to the cartillage, even to separate the diaphragm to the extent of an inch in the direction upward, and then, without opening the peritoneum, pericardium, or pleura, to cut through and remove the cartilages if diseased.

A meeting being held of the consulting surgeons of the hospital, it was decided that it was proper to perform the operation, and proceed in as far as the patient's safety would permit. The operation was performed on March 9th in the following manner: The patient being placed on a table, an incision seven inches in length was made from the fourth rib downward, and the anterior face of the tumor exposed by dissecting away the integuments with the externus oblique and rectus muscles, so far as these were not incorporated in the tumor. Its face being exposed, presented a bluish color, and was of a scirrhous hardness. Every stroke of the knife was followed by a copious flow of blood. When the circumference of the tumor was uncovered its edges were found to be quite undefined and concealed by the muscles above mentioned. These being cut through, an ill-defined edge was discovered, and the dissection was continued along the ribs, from which it was perceived that the tumor could be detached, although strongly adherent. When the dissection was carried as far as the edge of the ribs the tumor was found to turn down over the cartilages into the epigastic region, to involve

the internal oblique and transversalis muscles, and to adhere
to the peritoneum for the space of about an inch. From this
it was dissected up, and the whole tumor removed in a mass.
The latter part of the operation was much obscured by the
quantity of blood which was given out by the arteries on all
sides. Four large arteries and some smaller ones required
ligatures. The patient suffered much when the tumor was
raised, from its drawing the peritoneum outward. This part
of the operation, however, was short, and as soon as it was
terminated he ceased to suffer severely. The wound was
closed, leaving an outlet for the sanguineous oozing (sixth
and seventh ribs were resected).

On examination of the tumor it presented a cartilaginous
hardness. Its surface on all sides was composed of the
muscles between which it lay. Its substance consisted of a
brownish texture, in which a multitude of granulations, the
sixteenth of an inch in size, presented. At one point there
was a softening, as if suppuration was about to commence.
At another a discolored spot was seen. The internal or
epigastric part was equally hard with the rest of the tumor.
The surfaces of the cartilages were deeply depressed where
the tumor had lain.

The patient since the operation has had a smart fever, and
some appearances of peritoneal inflammation, which was
relieved by two or three bleedings. From this he is now
convalescent.

(Recurrence took place, but I can find no details.)

CASE IV.

Second case of J. C. Warren, 1837, osteosarcoma (*Boston
Medical and Surgical Journal,* 1837, vol. xvi. pp. 203,
204).

The patient was a stout, healthy person, a surveyor, aged
thirty years, from Upper Canada. Six years since, and without any previous injury that he was aware of, a small, hard
tumor appeared over the angle of the ninth or tenth rib.
From that time to the present it has gradually increased in

size—having, however, daily, less of that bony hardness which characterized it at first.

It was of a circular form, about six inches in diameter, and having an elevation above the ribs of between two and three inches. Its situation was on the lower part of the chest, covering a portion of the seventh, eight, ninth, and tenth ribs —to all of which it appeared attached, but more firmly to the ninth, which was most affected by its movements. The skin covering the tumor was perfectly natural, both in color and consistence, and the patient complained of no inconvenience or suffering from the disease.

The operation was performed on March 27, 1837, in the following manner : The patient was placed on a table a little inclined to the right, with a pillow under the chest, so as to cause a projection of the left side of the thorax.

The operation was commenced by a longitudinal incision, about five inches in length, directly over the tumor. This was joined by another transverse incision at right angles with its central part—the two incisions being well represented by the letter T. The flap thus formed being dissected up, the insertions of the external oblique were exposed and dissected off, not without some difficulty, however, on account of its strong contractions. The latissimus dorsi was now discovered passing over the outer edge of the tumor, and the same, in fact more, trouble was experienced in dividing this muscle than the preceding ; the dissection of it was excessively painful to the patient, and was followed by some hemorrhage. The tumor being at length perfectly exposed, was found to originate from the ninth rib, but was strongly adherent to the seventh, eighth, and ninth. A knife was now carefully insinuated under the tumor and its adhesions to the ribs dissected off, great care being taken not to cut through the intercostal muscle and penetrate the chest. As the attachment of the tumor to the ninth rib was not by its whole base, it was thought that more of the rib might be saved by first detaching the tumor and afterward cutting out the rib than by removing both tumor and rib together. It was, therefore, cut off from the rib at about an inch distance from its cartilage, and the morbid origin of the tumor exposed.

The intercostals were now cut through, the diaphragm carefully separated from the rib and pleura, and a director passed under at the points where the rib was to be divided. The bone was next cut through with the cutting forceps, and about two inches of it in length removed with a portion of its cartilage. The diaphragm immediately rose up, forming a hernia between the ribs. The hemorrhage was not great, most of it being from the divided muscles. No artery bled sufficiently to require being secured by ligature. The wound was brought together by sutures and adhesive plasters. The patient was very little exhausted by the operation. A considerable degree of febrile excitement followed, requiring the employment of copious bleeding.

The wound was united nearly throughout by first intention, and the patient is rapidly gaining strength without the occurrence of any bad symptoms.

In this case, unlike the preceding, very little thickening of the parts lying under the ribs had taken place. This rendered the difficulty much greater in separating the pleura and diaphragm from their adhesions to the ribs, which, however, was finally accomplished without penetrating the chest. (Quénu and Longuet mention resection of the sixth, seventh, eighth, ninth, and tenth ribs.)

Both of these operations sufficiently demonstrate that the excision of the ribs, with sufficient precaution, may be practised with comparative safety.

CASE V.

Case of Paget, 1853, chondroma (Paget's *Surgical Pathology*).

(Case No. 21 in Schläpfer von Speicher's list.) A man, age not given. The tumor was a pure enchondroma, formed of hyaline cartilage. It was removed, but recurred, and death resulted. The second tumor was mixed in structure, probably medullary sarcoma.

CASE VI.

Case of Gibson, 1853, chondroma. No. 26 in Schläpfer von Speicher's list. This was a boy, aged six years. The tumor seems to have developed as the result of a blow. The ribs and spinous apophyses were involved. The tumor was extirpated, but the boy died six years afterward from recurrence.

CASE VII.

Case of Foucher, 1859, chondroma (*Monit. d. sc. med. et pharm.*, Paris, 1859, vii., 126, 133. Schläpfer von Speicher's No. 15).

Man, aged thirty-five years (peasant). Right side, region of the breast. Five years ago he noticed a tumor the size of a hazel-nut. Stationary four years. Six months ago grew rapidly, with severe lancinating pains. On admission into the hospital it was the size of a large hen's egg. Skin normal, not adherent at any place. Very hard, not uniform, bossed; in shape a flattened ovoid. Axillary glands not affected.

Operation, November 13, 1859: Oval, circumscribing incision. Sprang from perichondrium of the fifth rib. Chill next day. Death two days later from purulent pleurisy. Autopsy showed right pleural cavity filled with purulent secretion.

Broca examined the tumor microscopically. Thought it mostly formed of cartilaginous tissue, having undergone some alterations, *but not cancerous*. Enchondroma not well understood clinically in those days, especially in those regions. Velpeau does not mention, in his *Traité des Mal. de Sein.*, enchondroma. The diagnosis should lie between scirrhus and enchondroma.

CASE VIII.

Case of Virchow, 1863, chondroma. Case No. 9 in the list of Schläpfer von Speicher. (See also Virchow's *Lehre v. den. Gesch.*, 1863, Band i.)

The patient was an old man, age not mentioned. It is not

stated where the tumor was situated nor on which side. It appeared to come from the ribs, but really came from the intercostal spaces. The tumor could not be completely removed; it was only cut off to the level of the thorax. The wound healed, notwithstanding an attack of erysipelas and pleural metastasis, which invaded also the daiphragm and lungs. The tumor was cartilaginous in structure, but had undergone ossification in parts.

CASE IX.

Case of Hueter and Langenbeck, 1865, sarco-chondroma. Taken from No. 4 in Schläpfer von Speicher's list. (See also *Archiv für klin. Chir.*, 1866.)

H. L., aged thirty-three years, letter-carrier. Tumor to the right side of the vertebral column. Admitted to the surgical clinic, Berlin, December 9, 1846. One year before a small tumor was discovered on the back, near the tenth rib. He had been struck by a heavy package falling from a post-wagon. The tumor grew slowly at first, without any severe pain. Later severe stretching and drawing pains came on, and the tumor began to grow more rapidly.

Examination. The tumor was found to be the size of a child's head, sharply limited, extending from the posterior inferior angle of the scapula downward to the lower edge of the ribs. The skin was movable, and could be picked up from the tumor. It could not be determined whether the pleura and peritoneum were involved. The matter was fully laid before the patient, the dangers and uncertainties of the operation being fully explained. He insisted upon the operation, so Langenbeck did it.

Operation : A half-circular skin-flap was taken up and the latissimus dorsi was partially cut. Very great difficulty in freeing the parts of the tumor. The tenth and eleventh ribs were lost in the mass; these were cut in front and behind with Liston's scissors, and the tumor was removed after separation of the soft parts. An area of peritoneum, four inches square, was exposed. Three ligatures were used. The wound was closed by fourteen sutures. The flap had contracted so much during the operation that it was now just sufficient to cover the defect. For several days there was frequent pulse, but otherwise the condition was good. On the third day some stitches were cut to let out secretions that were distending the wound. The wound was irrigated with permanganate of potassium solution.

On December 23d there was a small sloughing of the flap, the other portion being united, so that erysipelas, which occurred later in the flap, extending upward to the neck, did not disturb the wound in its healing to any extent. Three weeks after the operation the patient left his bed, and on February 1, 1865, after complete healing of the wound, he was allowed to go home.

A letter to Hueter, nine months later from the patient, showed that no recurrence was visible up to that time.

Examination of the tumor macroscopically and microscopically showed enchondroma with probable sarcoma.

Case X.

Case of F. Busch, 1869, osteosarcoma (Langenbeck's *Archiv f. klin. Chir.*, 1872, xiii. p. 49).

Bertha M., aged twenty years. Had a tumor the size of an infant's head under the angle of the scapula. It was not adherent to the skin, and slightly movable against the ribs. It had been growing about six months. The steps of the operation are not described, but it is stated that the tumor had destroyed the ribs and had affected the pleura. There were also metastases in the lungs.

The patient died of shock. The operation was unfinished.

Case XI.

Case of Thiersch, 1871, chondroma. No. 24 in the list of Schläpfer von Speicher.

The patient, a man, age not stated. The tumor was situated on the right side in front, starting from the second and third ribs. The tumor was not completely removed, and mortification of the intrathoracic portion took place. There was a cavity the size of an egg communicating with the exterior by an orifice the size of a dollar. It was an enchondroma.

CASE XII.

Case of Billroth, 1871, chondroma.   Case No. 5 in the list of Schläpfer von Speicher (*Deutsch. Zeitsch. f. Chir.*, 1881).

Man, aged thirty-six years.   Always healthy formerly; he had remarked since January, 1870, a painless tumor on the ribs, immediately under the right clavicle.   No cause assignable.   It grew steadily, and had reached on October 20th a vertical measurement of five inches and a transverse of four inches, reaching from under the edge of the clavicle to the third rib, and from the right sternal edge to beyond the mammary line.   Skin unchanged.   Tumor hard and firm, immovably fixed on the ribs and covered by the pectoral muscle. Tolerably sharply limited, and without pulsation.   Aneurism excluded as well as mediastinal tumor.

Diagnosis: Chondroma or sarcoma of the ribs.   It was evident from its rate of growth that it must soon end the life of the patient by either growing externally and breaking down or internally, producing compression of the lung and great vessels or internal metastasis.   On the other hand, it was clear that the extirpation of the tumor was dangerous because of the difficulty of avoiding opening of the pleural sac, or because radical operation might demand it.   Billroth's sections in such cases had shown, not thickening of the pleura, as with periostitis of the ribs, but thinning.   It was desirable, therefore, to set up preliminary adhesive pleuritis to prevent pneumothorax.   One of the assistants suggested the use of a number of needles passed through the lung in the neighborhood of the tumor, but against this Prof. Billroth expressed the fear that owing to the movement of the lung the needles might give rise to abscesses.   It was decided, then, first to remove the tumor only to the level of the ribs, in the hope that the traumatic inflammation would cause peripheral adhesion or that it could be brought about by cauterization later on.

October 27th. Incision five inches long across the tumor; after dissection the tumor was cut away with a resection knife. It proved to be a colloidal chondroma, made up of many smaller masses, in places calcified. The sharp spoon was then used. On the third day the temperature rose to 40°, tumor sloughing. An opening into the cavity was demonstrated with a soft catheter. The wound of perforation had enlarged, but the patient's condition had distinctly improved. Pleural cavity washed every two hours with warm water and five to five hundred permanganate solution. The wound became as large as a two-thaler piece, and through this a distinct view of the cavity could be had. With each expiration the lung came well through the wound; in inspiration it fell back.

November 9th, erysipelas.

The percussion now showed important changes, dulness reaching to the sixth rib on November 10th. On account of the increasing embarrassment of breathing, Prof. Billroth used the Dieulafoy apparatus, and drew 1400 c.c. of yellowish-brown fluid. The local and general condition thereupon improved, but this only for a few days, as on November 14th another aspiration had to be done (1300 c.c.), on account of the filling of the chest. The improvement was again immediate, but only temporary, as the patient soon became unconscious, and died November 20th. The history of the patient seems to show that the result was brought about by erysipelas and the pleurisy consequent upon it. The autopsy showed both lungs containing air. There were no other tumor-masses on the ribs visible; in the anterior mediastinum, however, were some nodules of colloidal cartilage.

CASE XIII.

Case of Kappeler, 1878, sarco-chondroma myxomatosum. Schläpfer von Speicher's case No. 1 (*Deutsch. Zeitsch. f. Chir.*, 1881).

S. S., female, aged fifty-one years. Field laborer, unmarried, was admitted April 5, 1878, into the hospital. About twenty-five years ago she observed at the upper part of the right breast a hard swelling, the size of a hen's egg. It gave no pain, and did not prevent her from working. She mentioned no preceding injury. No family history of tumors. Three years ago there was in the same breast a painful tumor, which disappeared under poultices, leaving the original swelling. Last summer and fall, while working, she had felt at times pain in this breast, which disappeared with warmer weather. In the last four weeks she thinks the swelling has rapidly increased. In the region of the right mamma there is found a tumor the size of a child's head, which extends outward into the axilla, upward to three fingers' breadth of the clavicle, forward to within a hand's breadth of the sternum; below it hangs over, projecting three fingers' breadth above the arch of the ribs. Its lower limit is in the region of the sixth rib. The tumor shows above a round, not sharply limited, swelling, of firmer consistence than the rest of the tumor. No swelling of cervical or axillary glands. Skin over tumor is movable, and can be picked up. In the underpart of the tumor is found a 10 cm. long opening, which was made outside the hospital, giving exit to three litres of a dark brown, odorless, colored substance, and a number of spongy pieces of bone, the smallest of which was 12 mm. long and 8 mm. wide, the largest 24 mm. long and 20 mm. wide. From this opening there issued a foul, brown-colored fluid. The temperature was elevated, pulse very frequent—120 to 134. The appearance of the patient is not cachectic.

Operation, April 6th, under ether by Dr. Kappeler.

A very long incision, convex below, was carried across the wall of the tumor. The tumor was easily shelled out above, externally and internally, adhering, however, firmly over a large surface to the fourth rib. Here great difficulties were encountered in loosening up the tumor. After extirpation of the cyst-wall the fourth rib lay quite exposed and raw for a distance of 10 cm. This was, with hollow chisel and sharp spoon, so cut away that finally there remained only a rudiment, 4 mm. thick, which was removed with the small saw. The finger could now be thrust into the cavity, where above, as well as toward the sternum, other tumor masses could be felt. These presented themselves as small sacs, filled with thick, colloid substance, and were removed in turn, after incision, with finger and sharp spoon. After scraping out a large sac which extended behind the sternum it seemed that the right auricle could be felt directly under the finger, while behind the rhythmical movements of the lung were felt and seen. The bleeding was considerable, checked by ligature, acupressure, and Paquelin cautery. The great gap was for the greater part covered with skin. Drainage. Lister dressing.

Duration of operation one hour. Narcosis good. After the operation great paleness of face, small pulse, and great weakness.

Though the dressing was frequently changed, complete asepsis was not attained. The temperature, however, for eight days only was ever up to 38.5°, and after this was normal. A part of the edge of the wound sloughed, and some small pieces of bone came away.

19th. The carbolic dressing was discontinued on account of subacute carbolism.

23d. Out of bed.

June 2d. Nearly healed. Patient discharged.

July 9th. Back in hospital. Over the fifth rib and under the skin the tumor was the size of a small apple; also a small tumor in the cicatrix, superficially degenerated. Complains of cough, stabbing and boring pains in the chest. Poor appetite, little sleep, general condition not good. Evening temperature elevated.

13th. Extirpation of the recurrences. Resection of 10 cm. of fifth rib with upper tumor. It was now seen that the tumor had grown well into the chest cavity. Toward the median line it reached nearly to the middle of the sternum onto the pericardium; posteriorly it pushed the right lung against the

vertebral column, and, when with the finger and sharp spoon
the neoplasm was as far as possible removed, the gray-blue
lung was seen to form the posterior limit of a tumor cavity,
which was projected as large as a man's fist into the chest
cavity.   The bleeding was considerable, and hard to check
owing to the tearing out of the ligatures in the tissue. Cavity
filled with gauze and skin sutured.   Strict Listerian dressing.
She did badly for a time, and there seemed to be a communication between wound and lung, as she said she felt the wash-
ing-fluid go into the lung, and she coughed violently.

20th. Air-bubbles escaped from wound.   After this she
improved, and by August 1st there seemed to be no lung
communication.

August 17th. Recurrence in the form of a round, bluish-
colored swelling.

26th. Rapid progress of the growth.

September 1st. Resected.   Small nodules scattered over the
wound.

6th. Chloride of zinc application.

From now on icterus developed and increased.   Bile continued, however, to pass by stool, and was absent from urine ;
liver and spleen enlarged.   Other growths show in the wound.
More zinc chloride.

23d. Temperature 40°.   Liver-line down to umbilicus.

25th. Lung communication again.   No albumin nor bile
in urine.

October 8th. For nearly two weeks temperature not above
38°.   Continued use of chloride of zinc had little effect since
the growth enlarged.   She became more dyspnœic, and died
November 16th in collapse.

Autopsy showed right lung infiltrated with little tumors,
the large arterial branches being also filled with these tumor
masses.   Similar masses in left lung.

The tumor was an enchondroma myxomatosum.   Prof.
Roth, of Basel, who examined the tumor, thought the original
tumor was undoubtedly an enchondroma, and more likely so
from the fact that the metastases in the lungs represented

distinct enchondroma-emboli. The original tumor and the recurrences all showed similar structure—enchondroma with myxomatous combination. Haab, of Zürich, found in periphery undoubted smooth muscle fibres, and although he did not have the advantage of examining fresh tumor, as did Prof. Roth, still his examination seemed to justify him in saying that "the tissue character of the tumor was to be defined briefly as fibrosarcoma cartilagineum."

"Following Lücke," says Schläpfer, "the fibro-cartilaginous enchondroma is also clinically of interest because it furnishes the intermediate stage of sarcoma."

At all events, the tumor in question is a pregnant illustration that enchondroma is not an absolutely benign tumor, and evidences distinctly the capacity for forming metamorphoses and combinations, recurrences, and metastases. In order to include both ideas, may I not, says he, propose the name for the tumor described of sarco-enchondroma myxomatosum?

CASE XIV.

Case of König, 1882, myxosarcoma. From Paget's *Surgery of the Chest*, 1896, p. 178. See also Campe's *Thesis*.

This case, reported by Paget in his *Surgery of the Chest*, was operated on by König in 1882. The patient was a female, aged thirteen years. The growth had been noted after rheumatic fever a few weeks before the operation; was thought to be an abscess, and was punctured. General health was very feeble. The growth, about the size of a fist, was situated on the left side, far back over the eighth to the eleventh ribs. There was dulness on percussion, and loss of breath-sounds for about a hand's breadth all around it. The heart was not displaced. It was a round-celled myxosarcoma.

The operation showed extensive malignant disease, which was only scraped away so far as possible. There was rapid increase of the disease, and death three weeks after the operation. At the autopsy there were signs of septic absorption. There was a mass of disease in the pleural cavity the size of a man's head.

Case XV.

Zwicke's first case, 1882, osteosarcoma, *Charité Annalen,* 1882 (1884).

Case of a woman, aged thirty-four years. Tumor of the left side above the breast, globular in shape, of the size of the palm of the hand. It began growing when she was eighteen years of age, growing very slowly until a few months before the operation, when it became painful. The skin over the tumor was reddened.

Operation : Incision through the skin and pectoral muscle discovered a soft, friable, vascular tumor, reaching from the second to the fourth intercostal spaces. As the tumor grew into the thorax, only the lower portion was removed. Notwithstanding this, it is stated in the report that the wound healed with mild suppuration, and that the patient, examined later, was entirely well, no recurrence having taken place. This does not agree with the account of Quénu and Longuet, who state that pleural adhesions were recognized and that the operation remained unfinished. They give the same reference.

Case XVI.

Zwicke's second case, 1884, osteosarcoma, *Charité Annalen,* 1884 (1886).

This was a tailor's wife, aged thirty-two years (Quénu and Longuet give the case as that of a man). The tumor was of the size of a goose-egg, situated on the right side above the costal border. It had been growing four months. It was immovable.

Operation : This is not fully described, but it is stated that the tumor was loosely attached to the tenth rib, and the surface of the ribs in the neighborhood was rough. The diseased tissue was scraped away with a sharp curette until healthy bone was reached. The tumor consisted of granulation tissue and enormous giant-cells. The result is not given.

CASE XVII.

Helferich's second case, 1885, alveolar sarcoma. From Hermann Plitt's *Inaug. Diss.*, Berlin, 1890.

This was a male teacher, aged twenty-seven years. He was accustomed to lifting frequently the school benches. He suffered from scrofula since quite a child. While lifting benches two years previously he began to have pains in the chest. After awhile he noticed a swelling in the mammary line under the left nipple, about the size of a cherry. When seen by Helferich, in 1885, the tumor was of the size of a fist. It was a firm, nodular growth, attached to the fifth, sixth, and seventh ribs. There was no lymphatic involvement. The tumor was removed with resection of the sixth rib, the pleura not being opened, although adherent at one point. Curetting was also done and the wound was drained. There was recurrence eight months afterward in the form of two small nodules. I can find no account of the operation to remove these growths. The tumor was alveolar sarcoma. Plitt says "left side," but Paget credits it incorrectly to the "right" side. (Paget, *Surgery of the Chest*, p. 176.)

CASE XVIII.

Case of Riesmayer, 1886, osteosarcoma (*St. Louis Courier of Medicine*, 1886, xv. p. 513).

Miss B., aged nineteen years, born in Germany, noticed a year ago a small tumor on the right side over the sixth rib, growing gradually larger and becoming very painful. The tumor, excised twice, returned. On December 26, 1885, the patient came to the Polyclinic, when a smooth tumor the size of a hen's egg, bluish-red color, not ulcerating, was noticed a little outside of the axillary line, connected with the sixth rib. Skin immovable over tumor.

Diagnosis: Osteosarcoma. Tumor excised. Tissues at some places removed down to pleura, and bone curetted. Wound left open, dressed with styptic cotton, and later with iodoform. The day after the operation patient complained of incontinence of urine and constriction around the abdomen, and was unable to move her legs; sensibility in lower extremities diminished. The paraplegia may be explained either by a myelitis or the formation of metastases. Died a few days later.

CASE XIX.

Case of Polaillon, 1889, sarcoma or chondroma? From his work *Affections Chir. du Tronc*, Paris, 1896, p. 203; obs. 104.

B. (Eugene), aged forty years, brickmaker, was sent to me at La Pitié, December 23, 1889. Following a contusion he had observed forming on the anterior extremity of the tenth rib, right side, at the level of its articulation with the cartilage, a tumor, which had enlarged quite readily and given rise to some pain. The pains radiated along the course of the intercostal nerves. The tumor was of the size of a green nut, hard, slightly bossed, skin unchanged in color, and freely movable. It was incorporated with the rib. It was either a sarcoma or an enchondroma. There was a formal indication for excising it.

December 30, 1889. Under chloroform, longitudinal incision, excising of the tumor, and scraping of the rib and cartilage at point of implantation. Listerian dressing. Almost immediate union, with very slight suppuration. Discharged January 14, 1890.

CASE XX.

Alsberg's third case, 1892, secondary carcinoma. From Caro, *Deutsche med. Wochenschrift*, 1893, p. 57.

Female, aged fifty-five years. Four years before admission the left mamma had been removed for carcinoma and twice in 1892 small recurrences had been removed. In October, 1892, she came back with a recurrence over the sternal ends of the fourth and fifth ribs. This was the size of the palm of the hand. The fourth and fifth costal cartilages and 3 cm. of the fourth and fifth ribs were removed. There was a tear in the pericardium 2 cm. in diameter. This was closed when the flap was brought down to cover the wound; the pleura was not opened. The wound was sutured, and after three weeks it had healed and the patient was doing well.

CASE XXI.

Bardeleben's second case, 1892, periosteal sarcoma (*Bericht u. d. chir. Kl. des. Geh.*, Prof. D. von Bardeleben *für die zeit*, vom. i. April, 1892, bis märz, 1893, von Stabartz Dr. Tilman *Charité Annalen*, xix., Berlin, 1894, 320 and 321).

This was a man, aged twenty-one years, bookbinder. He had suffered since childhood from a discharge from the ear. In November, 1891, he suffered from intermittent pains in the right side. At the end of 1891 there was a gradually increasing swelling of the right side of the chest until, in June, 1892, it had reached the size of a child's head, reaching from the fourth rib nearly to the arch below, and posteriorly, to the angle of the scapula. Measured with calipers, the length was 17.5 cm., breadth 10 cm. It was adherent to the chest wall. The tumor was elastic and the skin was movable over it.

Operation : The incision was in the form of a " T." The seventh and eighth ribs were involved and partly destroyed. The sixth, seventh, and eighth intercostal spaces up to the vertebral column were filled with tumor masses. The sixth and ninth ribs were only overlaid by the tumor. Radical extirpation was not found possible. The superficial mass only could be removed. Shortly after the operation there was collapse. Four hours later the temperature was 35.3°; pulse 152, very small. Camphor and ether injections, subcutaneous infusion of 190 grammes of salt solution ; autotransfusion. Coffee and sherry were tried without avail. Exitus letalis ten hours after the operation. Autopsy shows universal anæmia.

CASE XXII.

Bardeleben's third case, 1892, osteosarcoma. For reference, see second case.

This patient was a laborer, aged forty-nine years. He suffered in 1890 of sarcoma of the right chest wall. This disease was removed with partial resection of the seventh, eighth, ninth, and tenth ribs, and a large part of the co-tal arch, as well as thumb-sized projection reaching to the liver, which was tied off. The healing had been very slow; the patient was finally discharged with a fistula. In 1892, when coming under von Bardeleben's observation, this fistula still existed. In addition, a doughy tumor was seen from the third to the seventh ribs. The fistula seemed to go directly into the tumor. The fistula was split for 10 cm. and was found surrounded for 4 cm. of its length by new bone in tubular form. The fistula led to a necrotic point on the posterior surface of the sternum, at the beginning of the xiphoid process. This disease was removed with the chisel, and the soft mass scraped out with a sharp spoon and washed out with benzoated tincture. Moss dressing. Healed only after five months. Manifestly here was some remnant of the tumor, and it is remarkable that in one and a quarter years neither metastasis nor extension of the growth had occurred. The liver showed no special enlargement in spite of the fact that the tumor seemed to come from it. The upper edge corresponded to the seventh rib; the lower edge projected 3 cm. below the ribs in the parasternal line.

One and a half years later, at the time of this report, the patient felt entirely well. The cicatrix was loosely attached to the subcutaneous tissue; liver not enlarged, and no disease elsewhere.

Case XXIII.

König's fourth case, 1893, osteosarcoma (round-celled). Paget's *Surgery of the Chest*, 1896, p. 180, No. 22; see also Quénu and Longuet (*op. cit.*), p. 393, Case No. 10.

This was operated on in 1893. Female, aged thirty years. The growth had been noted a year; it had been of slow growth and without pain. It was a hard tumor the size of a hen's egg, lying from the sixth to the eighth left cartilages, just where they join the sternum. It was a round-celled sarcoma. Subperiosteal resection of the sixth, seventh, and eighth ribs, and cartilages, about two and one-half inches of each. A piece of the wall the size of the hand was removed. The growth had invaded the mediastinum, but was removed without opening the pleura. The pulsation of the right ventricle was clearly seen. She recovered in three weeks. There was gangrene of the attached skin over one end of the resected cartilages. There was no recurrence nearly two years later.

Case XXIV.

First case of Quénu and Longuet, 1896, osteosarcoma (*Revue de Chirurgie*, May 10, 1898, p. 400).

Man, aged fifty-two years, admitted August 13, 1895. Began in 1892 as a small, indolent tumor on the posterior aspect of the chest. This was extirpated in the service of M. Polaillon. The patient left the hospital incompletely cured, the wound being still unhealed. A month later recurrence was already manifested, but he would not immediately consent to operation.

In August, 1895, when he came into our service, the tumor was of the size of an orange, fairly well rounded, seated below the angle of the scapula. At its summit was an ulceration from which flowed a grayish detritus. There was occasional oozing of blood. No axillary involvement. The base of the neoplasm was not manifestly incorporated with the ribs. The second extirpation was now undertaken. This operation went to the periosteum of the sixth rib, which appeared only superficially involved. A fragment of this rib was resected. The other ribs were sound. The wound, which was closed with difficulty, showed a second recurrence a month later. Examination before the third operation showed a mass of the size of the fist, situated behind and below the left shoulder blade, ulcerated at its summit, expanded, and smooth at its base, which was incorporated with the skeleton. The skin was red and thin. No mobility, no pain. General condition good.

Operation, November 15, 1896. After having separated the tumor from the surrounding parts, not without some difficulty, a series of small resections all about the mass were made with the end in view of mobilizing a large portion of the thoracic wall, on which the tumor was seated. Thereupon the tumor was separated under the ribs by the method of pleuro-parietal stripping. The pleura, being thick, was quite easily denuded. Through it the movements of the lung could be perceived. The wound could not be completely covered. A hiatus of the size of the palm of the hand remained. Immediate autoplasty was practised by fashioning a broad, quadrilateral flap, pedicled in the neighboring parts, twisted on its pedicle and sutured to the circumference of the wound to be filled in. Healing *per primam.*

Inoperable recurrence three months later. Death from cachexia four months later.

Case XXV.

Second case of Quénu and Longuet, 1896, osteochondroma (*Revue de Chirurgie,* May 10, 1898, p. 401).

Man, aged nineteen years, entered October 19, 1896. Began three years ago as a small nodosity of the size of a hazelnut, seated on the third rib, left side. It increased without pain to the size of a walnut, and became visible externally. It was for this deformity he came to be cured.

On examination the tumor was of the volume indicated. Amalgamated with the anterior extremity of the third rib. Was fairly well rounded, of elastic consistence, and without pain.

Operation October 23, 1896. The tumor was dissected out, the fibres of the great pectoral muscle being easily separated. The third costal cartilage was denuded with the rugine, likewise the anterior extremity of the third rib, 5 cm. of this chondro-costal fragment being resected. This had served as a support for the neoplasm. The pectoral fibres were sutured, and the wound closed by superficial cutaneous sutures. The movements of the lung and heart could be plainly seen through the thin and supple pleura.

Union per primam. Six months later no recurrence.

Examination of the specimen showed a cartilaginous consistence; the implantation appeared to be at the union of the cartilage and rib, but it was hard to determine whether one had to do with an enchondrose or an enchondrome. The histological examination demonstrated its origin from the rib and its purely cartilaginous structure—that is, of hyaline tissue with very large embryonal cartilage cells.

CASE XXVI.

Case of Coley, 1896, osteochondroma of rib (*Annals of Surgery*, xxiv. p. 52).

" Dr. Coley presented an osteo-cartilaginous tumor, weighing two pounds, which had been successfully removed by operation from the right chest wall of a man aged twenty-five years. Since birth the man had similar small growths on one humerus and both tibiæ. Four years ago was first noticed a small tumor in the right side of his chest about the region of the nipple. This was from the first firmly attached to the rib, and appeared to be similar to the congenital tumors. It slowly increased in size until it became a mass about the size of a cocoanut, extending from near the middle line to the anterior axillary line, and from a point two inches below the clavicle to the free borders of the ribs. The mass was markedly protuberant, nodulated, and firmly adherent to the chest wall. The skin over the tumor was normal ; the general health was unimpaired.

" In the operation for its removal the pleural cavity was not opened ; there was very little loss of blood, and the patient bore the operation well." [1]

SERIES II. CASES IN WHICH THE PLEURA WAS OPENED, EITHER ACCIDENTALLY OR DELIBERATELY.

CASE I.

Richéraud's case, 1818; carcinoma. (Account of a resection of the ribs and the pleura, by M. Le Chevalier Richérand. Read before the Royal Academy of Sciences of the

---

[1] Three cases in Quénu and Longuet's list: two of Demarquay (fibro-sarcoma) and one of Morell-Lavallée (chondroma) are omitted because in none of them was the chest-wall involved.

Institute of France. English translation by Thomas Wilson, 24 pp., 8vo. Philadelphia, Town and others, 1818.)

GENTLEMEN : I have the honor to inform you of a surgical operation of which the record of the art furnishes no example—a new operation demanded by necessity and justified by success.

M. Michellan, surgeon at Nemours, for three years had been afflicted with a cancerous tumor in the region of the heart, the eradication of which a neighboring surgeon attempted in January last. At the removal of the first dressing a bloody fungus appeared in the centre of the wound; cauterized at each dressing, it grew again with activity. A second operation was attempted. This penetrated more deeply. After having laid the ribs bare he went to the pleura; in the meantime new fungi displayed themselves, and were reproduced, notwithstanding repeated cauterizations, by means of which it was endeavored to repress them. Grieved at not reaping any benefit from so many and so painful operations, the patient came to Paris toward the end of March, fully determined to suffer everything in hopes of being delivered from so dreadful a disorder, and of escaping an inevitable death.

At this period an enormous fungus was springing up in the wound. From this brown and flimsy vegetation oozed an abundant sanies, reddish, and so fœtid that it was impossible to remain a quarter of an hour near the patient without renewing the air of the apartment. The pains, nevertheless, were moderate; he experienced neither sweats nor colliquative diarrhœa; and, although tormented with an old and habitual cough, the patient, aged forty years, of a robust complexion, presented the most encouraging moral disposition. In this state of things it was resolved to attempt a resection of those ribs from which it was thought the cancer originated. Entrusted with this operation, I informed the patient that very probably I should be obliged to cut away a portion of the pleura; he did not hesitate to submit himself to that operation, all the consequences of which we did not conceal, and which he was able to appreciate.

All things being ready, I proceeded thereto on March 31st, encouraged in this bold undertaking by the enlightened as well as active assistance of my colleague, Prof. Dupuytren, and by other persons of the art who were so good as to lend me their co-operation. The patient presented himself to the instrument, refusing to be held by the surgical aids, and promising a firmness which did not belie itself.

I began by enlarging the wound, giving it a crucial form. I thus uncovered the sixth rib, which appeared to be inflated and uneven in about four inches of its length ; with a buttoned bistoury, the point of which I conducted along the upper and lower rims, I cut the intercostal muscles; afterward with a small saw, whose denticulated edge was only fifteen lines in length, I sawed the bone at the two extremities of the part affected.

This being done, I detached from the pleura the fragment thus isolated by employing therefor a simple spatula. I found in this an unexpected facility—a facility which proceeded from the condensation of the pleura beneath the bone, as the sequel of the operation has proven.

The seventh rib was laid bare for the same extent, isolated and detached in the same manner, but with much more difficulty and not without a slight tearing. The pleura discovered itself then, evidently diseased, thick, fungous, and giving birth to the vegetation in the space of the portion of the ribs taken away. The cancerous state extended itself above the sixth rib, so that the membrane appeared affected in about eight square inches of its extent. To make no excision of it was to leave incomplete an operation which lasted for twenty minutes, and until the present moment successful. Each of the assistants prepared himself with something capable of stopping the excessive hemorrhage, which was to be dreaded at the moment when I should make the section of the intercostal arteries. I cut the pleura with scissors whose blades

were crooked on the side of the edge ; and either that the section operated by this instrument—which cuts less in sawing
than in pressing, and bruises the tissue which it divides—
had determined the retraction of the vessels, or that their
size had diminished in consequence of the antecedent cauterizations, there did not run a drop of blood, but at this moment
the exterior air made an eruption into the chest, rushing in
with violence and compressing the left lung, which, with the
heart enveloped in the pericardium, was borne toward the
orifice.   I sought, by placing thereto the left hand, to moderate the entrance of air and to prevent suffocation, which
appeared imminent, while with the right hand I applied on
the wound a large bolster spread with cerate.   The entrance
of the air was suddenly stopped by this greasy cloth, large
enough to cover not only the wound, but moreover all the
side of the chest corresponding.   I fixed above a large and
thick piece of lint; I covered it again with some bolsters,
and supported all the apparatus with a rolled bandage, mid-
dling tight.

The anxiety and difficulty of respiration were extreme
during the twelve hours which succeeded the operation.   The
patient passed the night entirely in a sitting position.   Toward
morning sinapisms, applied to the soles of the feet and to the
inner surface of the thighs, rendered the respiration more
easy.   From that moment the pulse grew higher and the
strength was reanimated.   The patient took nothing else for
his drink and diet than an infusion of flowers of the linden
tree, and of violets sprinkled with some drops of water dis-
tilled from orange flowers and sweetened with the syrup of
gum arabic.   Three days passed thus : The fever was abated,
and the oppression sufficiently great to deprive the patient of
sleep.   The first dressing was removed ninety-six hours after
the operation.   The pericardium and the lungs had contracted
adherence with the contour of the quadrilateral aperture, a
kind of window made in front of the heart.   Happily the
adhesion between the pericardium and lungs was not complete; for, from the sixth to the twelfth day, through want

of this adhesion, an abundant serum had run from the chest and gushed out at each dressing. The quantity of serum which ran by that in the space of twenty-four hours might be reckoned a half pint. On the fifteenth day this serum, produced by an inflammation of the surfaces, ceased to run, and on the eighteenth day the adhesion between the pericardium and the lungs was completed. The air thenceforward ceased to introduce itself through the wound, the patient could lie on his side, and his sleep and appetite were entirely restored.

The wound, although until then dressed with a greasy linen rag immediately applied to its surface, rapidly diminished and exhibited a better appearance. On the twenty-first day the greased linen rag was dispensed with, and the dressing was applied as for a simple wound. The surface was covered with fleshy pimples, which had sprung up from the lungs and the pericardium.

The patient, who had for some days made trial of his strength in a garden belonging to the house which he inhabited, could not resist the desire of riding in a carriage through the streets of the capital. Having experienced fatigue from a ride of five hours, in which he visited the School of Medicine and caused to be shown him the parts of his ribs and of his pleura deposited in the cabinets of that establishment, nothing could prevent him from departing on the twenty-seventh day after the operation, or from returning to his place of residence, where he arrived without accident, provided with a piece of boiled leather large enough to cover the cicatrix completely when perfected.

I suffered not the opportunity to escape which presented itself to prove anew the perfect insensibilty of the heart and pericardium. Nothing warned the patient of the touch of the fingers softly applied to the organs. In the state of life the pericardium in man is possessed of such a transparency that the heart is seen through this membrane as if it were under a glass bell, perfectly diaphanous; it is so much so that we might have been led to believe there were no envelopes. It is so far from being so that this complete transparency is found in the pericardium of dead bodies, and it appears to me that, in this point of view, this membrane may be compared to the glass of the eye, which becomes dim and obscure at the approach of death. A large aperture, with loss of substance in the contour of the chest, not being necessarily followed by suffocation, a bloody discharge, or by a mortal inflammation of the organs toward which the exterior air finds there a free access, it appears to me there could be effected, in a disease to which the patient would succumb, for example, a dropsy of the pericardium. I say there could be effected in front of the heart an aperture which would not only permit the water in which that organ is immersed to discharge itself, but also to effect a radical cure of the disorder by determining the adhesive inflammation of the surfaces, by processes similar to those used for the cure of the hydrocele.

The same operation would be justified in order to expose the lungs partially affected and in taking away some parts of them by binding them with ligatures. It will certainly be said that like enterprises are rash; but how many operations deemed impossible within the last fifty years obtain in our day the most brilliant and best attested success?

(M. Richérand requests his professional brethren to whom a patient not too much debilitated by age or disease, affected with the hydropericardium, should present himself, to address the same to himself if they do not prefer to attempt themselves the operation he proposes.)

I will not occupy much longer the time you have been pleased to grant me; it is for those among you who are particularly engaged in the improvement of surgery to apprise me whether, in the views I propose, I am not left to misconceive by a vain desire of improvement; it is to those to whom it belongs to judge whether the fact which I submit to their sagacity can contribute in some measure to the advancement of the science as well as the comfort of humanity.

April 28, 1818.

Note.—The patient had rapid recurrence, and died one month later.

Case II.

Case of Sédillot, 1855, sarcoma. Gerulanos, *Deutsch. Zeitsch. f. Chir.*, October, 1898.

Sédillot resected two ribs with corresponding part of pleura and a piece of the lung. Recovered. Nothing as to recurrence.

Case III.

Case of Schuh, 1861, chondroma. Leonhard Meyer, *Inaug. Diss.*, Erlangen, 1889.

This was a man, aged sixty years, suffering with a large enchondroma reaching from the third to the seventh ribs, and from the left sternal edge to the axillary line, left side. In attempting to peel out some process of the tumor with the finger in the third and fourth intercostal spaces the pleura was opened; with a whistling sound the air rushed in. The opening was quickly stopped with the finger, and the operator cut away the upper part of the tumor and laid the soft parts over the remainder of the tumor, uniting the wound, and put on a compress dressing of charpie, held with plaster strips. He died in two days from pneumothorax.

Weinlechner comments on this case: "To-day the lethal termination would, perhaps, be prevented by drainage of the thoracic cavity carried out under Listerian principles." (L. Meyer, *Inaug. Diss.*, 1889.)

Case IV.

First case of Heyfelder (J. F.), 1861, chondroma (*Beiträge zur genaueren Kenntniss der Thoraxgeschwülste, Berliner klin. Woch.*, 1868, v. 369).

This was a man whose age is not stated. The tumor was a costal enchondroma, thirteen inches in breadth, with a vertical measurement of eleven inches, situated on the right side. The tumor was indolent. The skin could be picked up over the tumor. The tumor was very slightly movable, moder-

ately soft and elastic, rounded in shape. It seemed from the
rapid growth to be a pseudoplasm(?). The patient had sustained a fracture of the seventh and eighth ribs one year
before. A tumor began before the wound healed, and grew
rapidly without pain. The man said if Heyfelder would not
remove it he would hunt until he found some one who would
do it.

Operation : The operation was begun by a semilunar incision over the under part of the tumor. The seventh and
eighth ribs, which supported the tumor, were resected, together with the involved pleura. The hemorrhage was quite
abundant. After the stopping of the hemorrhage the wound
was closed. He did well until the fifth day, when the temperature rose and the pulse became quickened. During the
night of the fifth day he died so quietly that it was scarcely
observed by the attendants.

Case V.

Second case of Heyfelder, 1861, chondroma. A tumor for
ten years, the size of a pigeon's egg, then suddenly growing
rapidly, in three months filling the space from vertebral
column to sternum, from clavicle to a point below not stated.
Died in the first week.

Case VI.

Third case of Heyfelder, 1861, chondroma. G. L., aged
thirty-three years ; ten years before suffered of a breast inflammation. The tumor had been growing since 1851. Right
side.

Operation, March 25, 1861 (?). Exposed two to three
inches of the fourth rib, which was carious. Pleura dissected loose. Died April 7th. At autopsy third and eighth
ribs were shown to be involved. A fist-sized tumor was
found involving the pleura at the fourth rib, attached to and
involving the lung.

He felt doubtful whether he could have done the operation
completely.

Case VII.

Case of Langenbeck, 1873, sarcoma (Reported by Israel, *Eighth German Congress for Surgery*, 1879, p. 45).

This case was operated on in the year 1873 by Langenbeck. A woman, aged thirty-two years, with a sarcoma in the left lateral thoracic region. The tumor was the size of two fists, and involved two ribs and the corresponding costal pleura. The wound was washed daily with carbolic water, and the patient ultimately died three months after operation from carbolic intoxication. The urine had become almost black, but the irrigations were persisted in. So little was known at that date about carbolic poisoning that the condition of the urine did not warn of the carbolic intoxication. This case was most likely a victim of antiseptic irrigation.

Case VIII.

Case of Israel, 1873, sarcoma (Langenbeck's *Archiv f. klin. Chir.*, 1887, Band xx. p. 26).

H. W., pale and suffering from stomach troubles, presented herself, showing an elastic tumor of mamma size, seated upon the left wall of the chest in the axillary line, having no connection with the quite healthy mamma. At the extirpation it was shown that the sixth rib passed into the tumor. The pleura was torn in two places; air and blood entered the pleural cavity; consecutive pleurisy, with foul secretion; discharge through the lung. There was obstinate vomiting of all ingesta. Death followed after three months from exhaustion. Chronic septicæmia (?). At the autopsy no metastases and no deep disease were found. The left pleura, about the size of a fist, was filled with thick clots, and there was marked anæmia in all organs.

This case is not found in Quénu and Longuet's table, but Gerulanos gives it as a separate case from Case VII., and Cases VII. and VIII. are described separately in the reference given. Plitt, besides, gives it as a distinct case. While

4

there are some points of resemblance, the internal evidence
seems to be in favor of its being a distinct case.

CASE IX.

Case of Cassidy, 1874, chondroma (*Canada Lancet*, 1874–
1875, vii., reported by Sylvester).

W. H., aged twenty-nine years, a laborer, came to the
Toronto Hospital as interne, January 25th, complaining of
pain and a hard tumor on the right side, situated on the sixth
and seventh ribs midway between the spine and sternum.
About six years ago he noticed a small, hard tumor, the size
of a pea, at first movable; kept growing until about six
weeks ago, when it became painful and sore. It had reached
now the size of a man's fist, hard and immovable, attached to
the sixth and seventh ribs. Diagnosed as an enchondroma,
and operation deemed advisable. The doctor made a horizontal incision over the tumor, dissected back over the margin
on each side and got well down; introduced fingers into the
wound to feel for pedicle and separate surrounding tissues.
His fingers accidentally entered the right pleural cavity. The
external air at once rushed in and completely filled the space,
compressing the lung. Did not interfere with operation;
with bone pliers the tumor was nearly all removed. Found
to be quite hollow, centre having undergone fatty degeneration. Hemorrhage very slight, wound soon closed up and
brought into apposition by plaster covering the whole surface.
He was discharged in six weeks; wound completely closed,
and respiration on both sides normal.

Treatment: First antiphlogistic and then tonic; good,
nourishing diet.

CASE X.

Case of Wattmann, 1878, chondroma (Albert's *Lehrbuch
der Chirurgie*).

In the earlier editions of Albert's *Lehrbuch*, in speaking of
the rashness of some surgeons in removing large chondromata, Albert refers to a case of Wattmann, related by Dumm-

reicher, in which Wattmann attempted to dissect the pleura from the posterior surface of a tumor, but was unfortunate enough to open the pleura, so that the patient died of pyopneumothorax.

This case is No. 25 in Schläpfer von Speicher's list, but he gets it from the same source, so that no additional information can be gotten from him. The case is not mentioned in the last edition (1898) of Albert's *Lehrbuch der Chirurgie*.

CASE XI.

Case of Kolaczek-Fischer, 1878, chondroma (Reported by Kolaczek in *Deutsche Gesellsch. f. Chir.*, viii. 2, x. p. 80). This is a particularly interesting case, as it was on the border line between the pre-antiseptic and the antiseptic era. The successful result in this case gave a great impetus to the surgery of the tumors of the thoracic wall and to surgery of the chest in general.

It was operated on in the year 1878 by Fischer, and the patient was presented by Kolaczek at the meeting of the Eighth German Surgical Congress, in April, 1879.

C. S., female, aged forty-eight years. She had an enormous tumor of the left side of the chest, extending from the clavicle to the costal arch. It seemed to have originated from the fourth rib, and had a history of four years' development at the time of operation.

The operation required resection of parts of the fourth, fifth, sixth, and seventh ribs, and the subjacent pleura. The defect in the pleura was of the size of a child's head. There was at first powerful expiratory expansion and bulging forward of the lung, which thereupon immediately collapsed. Respiration became very anxious and the pulse failed. The dyspnœa for a time was intense, and the narcosis had to be suspended. The operation was rapidly finished and the patient's condition improved. The wound was sutured, provision being made for drainage. A salicylic acid solution was used for washing out the pleural cavity. Following the operation there was moderate dyspnœa, weakness, and anxiety, and the fever rose on the third day to 39.8°. There was a suppurative pleurisy and bronchitis, which Maas and others attributed to the antiseptic irrigation. By the third day the markedly drawn-in skin had formed adhesions with the lung and the pericardium. She made a good recovery, and was discharged four weeks after operation. There was recurrence one year later, which Kolaczek removed by resection of 3 cm. of the third rib, the tumor being the size of a walnut. She died August 18, 1881, in the hospital at Munich.

There was fungous disease of the ankle-joint and caries of the bones of the foot, for which she had declined all operation. The lung was found adherent all around to the thorax and only 1½ cm. in thickness. This case is shown in Figs. 1, 2, and 3.

This woman was the patient on whom von Ziemssen made some valuable investigations concerning the physiology of the heart. She had been examined in various hospitals in Germany, and was finally sent to Munich to see von Ziemssen. She was in Munich in August, 1879, and again in the fall of 1880. Von Ziemssen saw her in 1880. There was a defect in the chest wall measuring 9 cm. by 11 cm., bounded by the second rib above and the seventh below. The defect had a depth of 9 cm. The skin being drawn in, the lung lay along the vertebral column from its apex to the diaphragm; the nipple was well drawn into the cavity. The observations of von Ziemssen can be found in his studies on the " Movements of the Human Heart of Catharina Seraphin," published in the *Deutsche Archiv f. klin. Med.*, 1882, Band xxx. p. 270.

CASE XII.

Tietze's first case, 1879, giant-celled sarcoma (*Deutsch Zeitschrift fur Chir.*, 1891, Band xxxii. Reported from Fischer's clinic).

J. G., aged forty-six years, coachman, admitted to the clinic July 26, 1879. There was a history of injury, being kicked by the hind foot of a horse three years before. The tumor was of the size of a child's head, spherical in shape, situated on the right side of the chest between the parasternal line and the anterior axillary line. The tumor was immovably adherent to the sixth, seventh, and eighth ribs, non-fluctuating, but elastic. The skin over it was smooth, but showed venous turgescence.

Operation : Vertical incision. In the attempt to separate the tumor at its base the seventh rib was broken and considerable hemorrhage ensued. About 10 cm. each of the sixth, seventh, eighth, and ninth ribs were resected. There was adhesion of the diaphragm to the tumor, and in separating it the pleura was torn. On account of the adhesions the lung did not collapse, and the dyspnœa was slight. The loss of blood was marked, and the subsequent dyspnœic attacks were severe. The wound was sutured and drained. On the second day there was œdema of the lung and hemorrhage into the pleura and pericardium and endocarditis. Death on the second day. These latter notes were findings of the post-mortem.

CASE XIII.

Leisrinck's case, 1880, sarcoma (round-celled) (Langenbeck's *Archiv*, Band xxvi. ; Plitt's *Inaug. Dissert.*, Berlin, 1890).

J. S., aged thirty-seven years, had first noticed the disease only four months before, after a blow two months previously, which he thought had fractured a rib. There was a large tumor of the ribs, right side, extending from the fifth rib down to the lower border of the thorax, and from the edge of the sternum to the lower angle of the scapula. He was a strong, fine-looking man.

Operation: The incision was oval; extended from the fifth costosternal junction down to the costal arch, and thence up to the inferior angle of the scapula. The sixth and seventh ribs were resected from near the sternum backward for nearly their whole length. The pleura was torn, and, as the growth had begun to attack the diaphragm, a piece of the diaphragm was removed, about an inch and a half in diameter. Through the hole thus made the liver and some coils of intestine protruded, and had to be put back; the diaphragm was sutured. Instantly, when the pleura was opened, the patient collapsed, the pulse became very irregular and small, and the respiration embarrassed. Under ether injections (subcutaneously) and faradization of the phrenics he revived. The wound and pleural cavity were washed out with salicylic solution and sutured with two large drains. There was terrible cough during the night; next day respiration shallow, temperature 39.1° C. Thirty hours after operation lung sounds could be detected to the fourth rib. At first he improved, then dyspnœa came on, and the temperature mounted to 40.3° C. He died of suppurative capillary bronchitis, without infection of the pleural cavity, on the fifth day. The wound in the diaphragm was firmly closed.

Case XIV.

Weinlechner's case, 1880, myxochondro-sarcoma (*Wiener med. Woch.*, 1882, Nos. 20 and 21).

This was a man, aged thirty-seven years. Operated on in 1880. The tumor was the size of a man's head, involving the third to the fifth ribs of the right side. For the most part hard, but soft in places. It had been noted four years before, but after a blow seemed to have grown very rapidly. The third, fourth, and fifth ribs were resected with the involved pleura. The lung collapsed at once, and the condition of the patient was bad. The dyspnœa was severe, pulse small. A piece of the lung "the size of a saucer" was removed, and two small nodules from the upper lobe of the lung. When the pleura had been widely opened the patient seemed to be in the pangs of death, with very bad pulse, and the operation had to be hurriedly finished. The thymol spray was used during the operation. The wound could not be closed, as there were no flaps, so it was packed loosely with gauze and sponges. In the afternoon there was still dyspnœa; small, weak, accelerated pulse. Death after twenty-four hours from collapse and suppurative pleurisy (?).

CASE XV.

Kröulein's case, 1883–1888, sarcoma (*Deutsche Zeitsch. f. Chir.*, 1893, p. 37 ; Müller's article).

This was a woman, aged eighteen years. In June, 1883, she had a tumor on the left side which had begun eight months previously, and had reached the size of a child's head. This was connected with the sixth rib, which was resected and the tumor removed. Six months later a piece of the chest-wall, of the size of the palm of the hand, was removed, together with a growth in the lung of the size of a walnut. The space, resulting from the former operation, had been filled in, with a recurrence, which extended from the fifth to the seventh rib. The wound was sutured with catgut. The patient recovered, the wound healing in four weeks. In March, 1887, three years later, there was a recurrence above the old cicatrix. The fifth rib was resected and the thorax again widely opened and a piece of the lung excised; the lung collapsed and the patient became very cyanotic. The pleural cavity was washed out with 1 : 5000 sublimate solution. She returned in August, 1888, showing a recurrence in the region of the former operation. Further operation was declined. On November 29, 1890, she died of pneumonia.

CASE XVI.

Humbert's case, 1884, sarcoma (*Révue de Chirurgie,* 1884, vol. vi. p. 297).

L. B., aged twenty-one years. Operated on, February 12, 1883, by Dr. Peyrot, the diagnosis being periosteal sarcoma of the ribs. A second operation was done on December 5, 1883, by Dr. Nicaise. Patient again presented herself for treatment in December, 1884. The growth now involved the seventh, eighth, and ninth ribs, and was very painful.

Operation : Two vertical incisions in the form of a very elongated ellipse, which took in the old scar. Two other transverse incisions were made— one above, the other below. Two cutaneous flaps were thus thrown off, exposing the

tumor. It adhered to the seventh, eighth, and ninth ribs
and the adjacent intercostal spaces. The eighth rib, completely
involved in sarcomatous tissue, was fractured during the
manipulation. The three ribs were dissected clear and then
resected to the extent of 9 cm. each. He then made an incision along the sixth intercostal space and carried the knife
down through the points of section of the ribs in front and
behind. There was some escape of sero-purulent fluid. He
was now able to pull down a sort of trap-door, which was
adherent only below. On its internal surface the deep part
of the tumor showed in relief, well circumscribed in appearance. This opening disclosed a large cavity lined with false
membranes. The lung, pressed backward, occupied the depth
of this cavity.

This thoracic flap was now grasped with the fingers of the
left hand and the scissors carried cautiously along the ninth
intercostal space, so as to be certain of passing beyond the
limits of the tumor. He had still to resect a small fragment
of the ninth rib, which seemed suspicious of malignancy. In
clearing the pleura he perceived on the floor of the pleural
cul-de-sac, near the inferior border of the thoracic opening,
a hole 2 cm. by 7 cm. in extent. Taking this to be an
accessory cavity, he introduced his fingers, but was not a
little surprised to come upon a smooth surface, limited below
by a sharp edge. This proved to be the liver, and he was in
the peritoneal cavity.

With the aid of a mirror he was able to see clearly this
viscus, the upper part of the ascending colon, and some intes-
tinal coils. It was evident that he had accidentally cut a part
of the diaphragm which adhered to the tumor. Examination
of the tumor showed small bundles of muscle-fibre, broken
down by the disease, but the peritoneal lining had as yet suffered no alteration. Some clots discovered on the surface of
the liver were removed by means of a mounted sponge.
Having made as well as possible the toilet of the peritoneal
cavity, he closed the diaphragmatic wound with five points
of strong catgut suture. The pleura having been cleared by

curettage of adhering false membranes, the wound was sutured, a drain being left at the inferior angle. An antiseptic dressing was applied. The patient was in a condition of pronounced algidity for twenty-four hours, and required hot applications. Recovery was without further incident, and she left the hospital February 6, 1884, less than two months after the operation.

In 1886 there was still no recurrence.

CASE XVII.[1]

Case of Helferich, 1885, sarcoma (Reported by R. Heigl, *Arbeiten aus dem med. klin., Institute der K. Ludwig-Maximilians-Universität zu München,* Band ii., 1 Hälfte, Leipzig, 1890, p. 405 ; also *Deutsche Archiv f. klin. Med.,* Leipzig, 1889, xlv. 27–42).

A. W., aged ten years, son of a peasant, of G. in Oberbäyern, was admitted into the stationary division of the K. Chir. Poliklinik at Munich, showing in the left pectoral region an oval-shaped tumor about the size of a man's fist, on all sides sharply bounded, covered with normal skin, and lying under the musculature. It extended from the under edge of the clavicle above to the upper border of the fifth rib below, and from the middle line of the sternum, internally, to the anterior axillary line, externally. Length, 12 cm.; breadth, 10 cm.; of firm consistence ; no pain ; no glandular involvement. Conformation of the ribs unaltered. Chest organs sound.

Diagnosis : Sarcoma. Nothing in the history, except that it was noticed first at the end of April.

Operation, May 30, 1885, by Prof. Helferich. The operation consisted in the cutting of a musculo-cutaneous flap and its reflection upward, freeing of the tumor, and resection of the fifth, fourth, third, and second ribs in the anterior axillary line, and of their insertions, with saving of the periosteum. At the second rib a large piece of the sternum could be saved, but a part of the manubrium had to be removed.

---

[1] This case has been improperly placed here ; should be included in Series I.

The resected parts of the ribs were bent inward and leveled. During this manœuvre nearly the whole pericardium and a large part of the costal pleura were exposed to view. The heart and vessels could be distinctly seen through the delicate pericardium, while the sharp edge of the lung could be studied during the respiratory and cardiac movements.

The healing of the wound, at the beginning disturbed by an enormous acceleration of the heart's action and dyspnœa, later became quite smooth. Microscopical examination of the tumor showed typical fibrosarcoma. The defect was protected by a soft pad.

Discharged July 13th. Over both lungs vesicular breathing, even in the defect. The effect of atmospheric pressure was almost absent, being apparent only on its inner edge. The regular rhythm of the heart could be well differentiated from the forward bulging caused by the air-filling lung, which was observed more laterally. Normal heart sounds; no pericardial friction.

In the summer of 1886 he was quite restored to health, and could be submitted to a more careful examination. There was no sign of recurrence, but there were some changes in the defect of the chest-wall, to be ascribed partly to physiological growth, partly to regenerative processes. Instead of an enlargement of the defect, as was to be expected, it had much diminished by the growth of the ribs and sternum. The sternum extended 2 cm. beyond the middle line above, and 4 cm. beyond at the ensiform cartilage. The axillary ends of the resected ribs reached now nearly to the mammary line, a firm, fibrous cord, extending from the second rib to the first costo-sternal joint. The fifth showed a conical movable process, running in a curved direction toward the sixth rib, somewhat pushed inward.

Measurements: Height, 134 cm.; chest measurements: in the axillary line, deep inspiration, 71.5 cm.; expiration, 66 cm.; ensiform measurements: inspiration, 73.5; expiration, 70. The heart pulsation fell during the long narcosis to 50 per minute.

Vou Ziemssen studied the Heigl case during the winter of 1888–1889.

The boy was now fourteen years of age—a strong, healthy fellow. The defect in the chest-wall was not so large as in the Seraphin case (Case XI.). It had the form of an irregular triangle, the base being below, the inner side formed by the sternal edge. Vertical measurements 11 cm., breadth 8 cm. The heart could be felt in its whole anterior extent. Both lungs freely movable, and the lung of the affected side seemed to be functioning over a greater extent than in the other case. On coughing and enforced expiration the lung and heart are driven into the defect. There was absolutely no sign of recurrence.

Case XVIII.

Case of Maas, 1885, chondroma (Langenbeck's *Archiv f. klin. Chir.*, 1886, xxxiii. p. 314).

H. M., aged forty-two years, peasant, admitted into Juliusspital, Würzburg, February 19, 1885.

Mother healthy. Father died, probably of rectal carcinoma. Himself uniformly healthy, married, seven healthy children. Fifteen or sixteen years ago, after a blow from a heavy stick of wood, developed a tumor of the size of a hen's egg under the left shoulder blade. The injury was not severe, and he did not quit his work. The tumor grew slowly until five or six years ago, when he fell with a sack of pears from a tree to the ground on to the sack, which lay between him and the ground. Now the tumor grew more rapidly and troubled him greatly at his work.

When admitted he was a large, well-built man, but somewhat emaciated, with pale skin and mucous membranes. Chest and abdominal organs healthy. A large, vertical oval tumor occupied the posterior and lateral aspects of the left side of the chest, reaching from the (by it) slightly upward displaced scapula at the seventh rib down to the crista ilei and anterior spinous process, and laterally from a point in front of the axillary line to within 3 cm. of the spines of the vertebræ. Greatest vertical measurements 41 cm., greatest breadth 37 cm. Skin normal except as to some dilated veins ; could be picked up over the tumor, except at its apex, where the skin was quite stretched by the underlying growth. Tumor elastic, of cartilaginous hardness in some places ; soft at the apex. The process reaching toward the pelvis was quite hard. Firmly fixed to the ribs.

Diagnosis : Chondroma or osteochondroma.

Operation, February 21, 1885. Under chloroform and a steam spray of acetate of alumina, patient lying on the right side (opposite) with a roll-cushion under him to raise the left thoracic and lumbar regions. Incision from inferior angle of scapula to left anterior spinous process, through skin and muscle. A second incision obliquely from the middle of the first, backward and downward through the skin and the back muscles. The exposed tumor, it could now be seen, was free above, and could be lifted from the ribs. It was firmly attached to the ninth, tenth, and eleventh ribs. The pelvic process had pushed its way downward between the musculature and the fascia transversalis to the anterior spinous process. Blunt dissection posteriorly, and cutting of the ribs with bone forceps made it now possible to lay the tumor forward. Now it could be seen that the pleura was firmly incorporated in front down to the diaphragmatic insertion. The pleura was cut first behind, and then the ninth, tenth, and eleventh ribs in the axillary line in front of it. There was nothing left to do but dissect up the lower projection of the tumor. This was accomplished without tearing the fascia or opening the peritoneum. Eleven cm. of the ninth, tenth, and eleventh ribs, with corresponding pleura, were resected. The defect in the wall was 11 cm. long and 7 cm. wide. The heart, covered with pericardium, the uninjured diaphragm, and the completely collapsed lung could be distinctly seen. The left kidney and the spleen could be seen shining through the transversalis fascia. The pulse sank from 80 to 60, and became small, but regular and easily felt, the breathing slow but regular, so that the narcosis did not have to be interrupted nor any restoratives applied. The patient was now laid on the left side, and the blood in the cavity, partly by gravity, partly by sponging, cleared out. The wound was closed by two rows of sutures, one uniting the musculature, the other the skin, only a small opening near the upper end of the longitudinal wound under the scapula and another near the hinder end of the transverse incision were left unsutured. No other drainage. Later there was slight dyspnœa. Sublimate salt dressing. The tumor was made up of a large cyst and a great number of smaller ones.

The microscopic finding was osteochondroma myxomatosum.

After return of consciousness the patient had wine, which brought up the pulse to 72. The respiration was little quickened, and very slight dyspnœa, without cyanosis. No cough nor expectoration. The course of healing was quite favorable and undisturbed.

February 26th. Examination showed almost normal respiratory movements, and no abnormal sounds. In inspiration the wall was drawn in only moderately. Deep inspirations, sitting up in bed, even standing upright, gave no pain whatever.

The dressing was thoroughly soaked the first day, so that it was changed on the third day. Wound completely adherent. Dressing changed again on ninth day. Wound healed except at the unsutured openings, which granulated. Temperature was once up to 38.5°. Patient got up on the ninth day, and went home March 14th.

June 16th. Returned for examination. Health good; respiratory functions practically normal; both lungs came down to same level. The defect had narrowed itself to 3 cm. in diameter. No pleural adhesion. A small, movable tumor above. According to Quénu and Longuet this was removed, and fourteen months later there was no recurrence. This case is shown in Figs. 4, 5 and 6.

Case XIX.

Schede's case, carcinoma secondary to carcinoma of the breast (*Deutsche med. Woch.*, September, 1886, vol. xii. p. 646).

Woman, age not given. In 1882 the patient came for operation on the breast. In 1885 there was a return of the disease, and a second operation was done. A portion of the thorax wall, the size of a saucer, was removed. The ribs were broken down by the growth. The pleura showed small growths (nature not stated), which were not removed, although the pleura seems to have been opened. The wound was covered by a skin-flap, which came into direct contact with the lung. This patient was exhibited a year afterward, and showed no evidence of a return of the disease. (These data are taken from a report of a number of cases by Schede at a meeting of the Hamburg Medical Society, May 18, 1886.)

Case XX.

Case of Trendelenburg, 1885, myxochondroma (Leonhard Meyer, *Inaug. Diss.*, Erlangen, 1889 ; Baldus, *Inaug. Diss.*, Bonn, 1887).

This was a man, aged sixty-one years. There was a tumor seated on the anterior aspect of the left thorax, semi-spherical in shape, and reaching from the sternum to the anterior axillary line, and from the clavicle to the mamma.

Diagnosis : Chrondromyxoma. The incision was made from the insertion of the fourth rib obliquely across the tumor to the axilla ; the flap was raised and the pectoralis major lifted, and the tumor uncovered in the direction of the external incision. Then a vertical cut was made perpendicularly to the middle of the first, reaching up to the clavicle. The pectoralis minor was cut across. The tumor was firmly fixed to the rib and evidently growing into the pleural cavity. It involved also the third and fourth cartilages, which were largely cut through. The tumor could now be lifted sufficiently to see the as yet uninjured pleura. In working downward and inward the pleura suddenly tore. The wound was immediately covered with a sponge. There was furious bleeding from an intercostal artery, which stopped the operation for a time. After a little manipulation of the tumor it was easily drawn out. The pleura could now be seen over an area the size of the palm of the hand. Sponge-pressure was kept on the pleural tear, and the median piece of the second and third ribs was resected. Then followed suture of the musculature and skin, the pleura-covering sponge being drawn out before completing the closure of the wound. Drainage through the external angle of the wound ; pressure bandage.

Discharged in three weeks, cured.

NOTE.—This seems to be the same case as that referred to incidentally in Witzel's account of his case (*Centralb. f. Chir.*, 1890, xvii. 523), in the following words : " The heart, however, here and in another former operation suffered more than was to be accounted for by the hampered respiration (this Witzel attributed to the flexion of the large vessels due to the dislodgement of the heart.) In the operation now referred to the tumor had become firmly attached to the middle lobe of the lung, considerably prolonging the procedure. The heart action was helped by the lateral support of a loosely rolled ball of iodoform gauze." Quénu and Longuet credit Witzel with two cases in their list. They give for both the same bibliographic reference as above, while in this reference, which is really Witzel's own article, Witzel mentions the case as one given in detail by Baldus in his *Inaug. Diss.* This corresponds in all respects to the Trendelenburg case now given.

CASE XXI.

Tietze's second case, 1886–1889, spindle-celled sarcoma (*Deutsch. Zeitsch. f. Chir.*, 1891, Band xxxii.).

R. H., aged sixty-six years, clergyman. In the fall of 1884 he fell and struck against a stump, bruising the right side. Shortly afterward considerable pain was felt, and in December of the same year a swelling was noticed close to the right nipple, increasing in size slowly, but with much pain. Entered the clinic in 1886, the growth being the size of a female breast; the skin over it movable, but tumor adherent to the ribs.

Tumor at operation was found to be a nodular mass of soft consistency in the centre. The fourth and fifth ribs were broken down by invasion of the disease. The tumor was removed close to the level of the ribs, and the ribs then resected. The adherent pleura, which was easily separated, was slightly torn, but was immediately clamped and sutured. A drain was put in and the wound sutured. The patient was discharged after the fourteenth day.

July, 1887. A small recurrence, immediately removed by scraping.

In December of the same year a second recurrence. While under treatment gangrene of right foot developed, necessitating amputation at ankle. Patient discharged well in January, 1888.

April, 1888, a third recurrence at the same site. This was removed by the fourth operation. The growth involved the third rib, extending up into the intercostal space. The rib was cut at some distance from the tumor and the attempt made to lift the tumor out; but the pleura was markedly adherent, so that a piece of the latter of the size of a 5-mark piece had to be removed. The lung did not collapse, and the pulse remained unaltered, as likewise the respiration. When the pleura was opened the chloroform was discontinued. Toward the end of the operation the breathing became stormy in character, the right lung pressing strongly against the defect in the chest wall and ballooning out. The wound opening was tamponed with iodoform gauze and then sutured.

The patient recovered rapidly from the operation, and seemed for a time quite comfortable. On the fifth day the tampon was taken out and a part of the sutures removed. The remainder of the sutures were removed on the third day. Discharged on the eighteenth day. The tumor showed spindle cells, with intermingling of round-cells in some parts.

February 28, 1889. Recurrence for the fourth time. During the operation the pleura was again opened. After apparent convalescence for a time, on the fourteenth day after the operation the patient showed marked nervous symptoms, became delirious, and died two days later. Iodoform poisoning was not considered probable.

CASE XXII.

L. Desguiu's case, 1887, sarcoma, spindle-celled (*Bulletin de l'Académie Royale de Méd.*, Belgique, 1887, iv. s. Tome ii. ; also *La Presse Médicale Belg.*, August 28, 1887).

This case was operated on in 1887. The patient was a man, aged forty-one years. A small tumor, seemingly inoffensive, developed slowly on the ninth rib and in its neighborhood on the right side. It began to be painful. There had been already two operations. Before the third operation it was found to be a sarcoma instead of a chondroma, and the tumor advanced rapidly in its growth. The sarcoma seemed to be of the spindle-celled variety. The eighth and ninth ribs of the right side, with adherent pleura to the extent of 4 cm. by 8 cm., were resected. Notwithstanding the ineviable pneumothorax, the healing was obtained by first intention, without a sign of pus and without any grave symptoms. The pneumothorax disappeared within a week, but the lung seemed only partially to expand, and the liver rose above the normal level. He recovered. The mildness of the manifestations on opening the pleural cavity were such as to surprise the operator.

The report was made four months after the operation. The Royal Academy of Belgium recognized the importance of the case and reproduced it in the report of its proceedings. It was remarked in this report that the case published by M. Desguin had no analogue for establishing the operation on a footing with pleural resection after the method of Estländer, more or less modified, having for its end the cure of pleural abscesses.

Case XXIII.

König's second case, 1887, myxochondro-sarcoma (From Gerulanos, and Paget's *Surgery of the Chest*, 1896).

This was operated on in 1887. Female, aged twenty-six years. She had broken a rib a year and a half previously. Six months later swelling was noticed at this point. There was no pain, but some tenderness. Below the right breast, over the sixth to the ninth ribs, was a large, flattened growth the size of a fist, giving a marked feeling of fluctuation. The eighth rib was resected with the adherent parietal and diaphragmatic layers of the pleura to the size of a thaler piece. The diaphragm was not perforated. The pleura, however, was freely opened. There was considerable dyspnœa for several days. Recovery took place in three weeks. The temperature rose to 39°, and was normal first by the twelfth day.

Recurrence three years later; very extensive; successfully removed by resection of the seventh and ninth ribs. There was invasion of the diaphragm. There was recurrence later (Campe).

Case XXIV.

Heincke's case, 1887–1889, sarcoma (Reported by Leonhard Meyer, *Inaug. Diss.*, Erlangen, 1889).

F. W., aged thirty-one years; married; parents living and well. Eleven sisters and brothers. One brother died of lung trouble at twenty-five years of age; one brother, now living, seventeen years of age, with a suppurating sore on the left leg. Always a healthy man until the winter of 1886–1887, when he had a cough without expectoration. He was the father of five healthy children. About Easter, 1887, he fell, striking heavily on the right side against a four-cornered timber. He suffered so much that he had frequently to see a doctor. There were pains in the side and painful breathing. At the end of September, 1887, he noticed a small kernel under the skin; by the end of October this had reached the size of an egg. He saw a physician, who advised operation, but he was fearful, and hesitated. The pain, however, became so severe and so kept him from sleeping at night that he came into Heineke's Erlangen clinic. This was December 10, 1887. The patient was middle-sized, not specially strongly built, but of good musculature and of moderately fat and pale skin.

Examination : The form of the thorax was rather paralytic ; the supraclavicular and infraclavicular spaces were sunk in ; clavicles were prominent, especially the left. There was slight flattening of the left apex. A tumor the size of a fist, raised about 4 cm., was seen at the lower part of the thorax. It was red in appearance, sharply bound, with an ovoid base extending from the scapular line to the posterior axillary line, occupying the region of the seventh, eighth, and ninth ribs, its longest diameter being on the eighth rib. Adherent to the skin and immovable on the ribs. Sharp limitation; no infiltration ; no enlargement of the glands. There was evidence of apical consolidation.

Operation, December 12, 1887. Skin incision not mentioned. There was resection of the seventh, eighth, and ninth ribs in the extent of the tumor. The pleura was then opened, and the piece on which rested the base of the tumor was cut out; the finger could feel a tumor the size of a pigeon's egg at the level of the eighth rib, nearer the vertebral column, projecting from the pleura inward. This was completely peeled out with the finger. The whole hand could now be put into the cavity. With the collapse of the lung a row of small tumors could be seen along the vertebral column. These, unfortunately, could not be removed. The pleura was sutured except at the posterior angle of the wound, where it was left open to the extent of 3 cm. for drainage. The external wound was closed except at the posterior angle. Moss pillow dressing.

Remarks : The opening of the pleura and the elimination of the right lung from the work of respiration gave rise to no disturbance of the narcosis. Naturally there was some dyspnœa after the patient awoke, but this was not particularly marked. There was, above all, no noteworthy heart disturbance. The tumor showed itself microscopically a round-celled sarcoma; the intercostal nerves were embedded in the tumor, an explanation of the pains. The day after the operation the temperature was 38.9°, afterward normal. He left his bed December 24th for a little while, but feeling chilly and badly, after half an hour he lay down again. He had a restless night. Temperature next day was 39.3°, lasting several days. On January 13, 1888, he was discharged at his request, the temperature being normal. There were no signs, on examination, to explain the cause of the trouble. On December 25th the surgeon had thrust in his finger at the open angle of the wound, thinking there might be some retained fluid, but only a little bloody serum came. The wound was closed, and when discharged it was completely healed. At home the wound suppurated again a little, but was healed again in the beginning of February. Some days after this healing, during a spell of coughing which lasted a quarter of an hour, a quart of foul, stinking, yellowish-green pus came up. These attacks were subsequently repeated, with less discharge. This continued, getting less and less through March, April, May, and June. He got stronger and felt better.

In July, 1888, he remarked a tumor at the site of operation, and returned to Heineke's clinic July 16, 1888. The skin was observed to draw in during inspiration, and ballooned out as large as a child's head during expiration.

Diagnosis : Recurrence of sarcoma in the cicatrix, increasing apical consolidation, and hernia pulmonalis at the site of extirpation.

On July 18th the tumor was extirpated through an incision in the cicatrix, the pleura being again opened and closed with suture. Union by first intention. A piece of the herniated lung was removed at this operation.

CASE XXV.

Müller's case, 1888, chondrosarcoma (Taken from Paget's *Surgery of the Chest*, p. 169. See also W. Müller, *Deutsche Zeitsch. f. klin. Chir.*, 1893, p. 41).

Paget does not include this in his table of cases, but gives it in full in the body of his work, p. 169.

A man, aged twenty-four years, in good general health, had a hard, rounded, flattened, shelving growth over the right ribs, from the fourth to the seventh, firmly united to them and extending from near the sternum to near the anterior axillary line. He had noted it four years. The veins over it were dilated, the skin was not adherent, the axillary glands were not enlarged. There was no dulness of the mediastinum, no displacement of the heart. Faint vesicular breathing was heard in the region of the growth ; no râles ; no friction sounds.

Operation, November 8, 1888: A large flap was raised, and it then could be made out that the intra-thoracic portion of the growth was much larger than the external swelling. Four inches of the fourth rib were removed, and then the fifth rib was similarly treated; at this moment the pleura, which was very thin, was torn, and air was heard entering the chest. Part of the sixth rib was now resected, pleura and all, and the tumor was raised, exposing a gap in the pleura as big as a saucer. At this moment the lung receded, and the growth went with it a little way back into the chest. Immediately the patient collapsed, his breathing stopped, his pulse could not be felt, injections of ether and camphor were given, and the danger passed off as soon as the growth was again grasped and drawn forward. It was now found to be firmly fixed by a broad base to the lower edge of the lung. The surgeon passed his finger over the lung and felt no other growths in it. His assistant then held the edge of the lung forward and kept it compressed while he transfixed and tied it, cut away the growth with scissors, and closed the wound in the lung with a continuous suture. He then let go the lung ; it collapsed at once; again the patient became deadly pale, and gave no sign of pulse or respiration ; he seized the lung and drew it forward, and again the patient revived. When, on tying his last skin suture, the surgeon had finally to let go the lung, the patient once more collapsed, and for some days after he had slight dyspnœa. Three days after the operation six or seven ounces of serum were let out from the pleura. In March, 1891 (two and a half years later), a recurrent nodule beneath the scar was removed. A little air entered the pleura, but no harm came of it. In July, 1893, the patient was in perfect health. The growth was of a mixed character, mostly bone and cartilage at its centre, sarcoma at its exterior.

Case XXVI.

Hahn's case, 1888, myxochondro-sarcoma (*Deutsch. med. Woch.*, 1888, No. 50, p. 1034; also the same, 1889, p. 321).

Riesenfeld refers to this case as having been related the day after the operation, before the Berlin Medical Society. The patient seems to have been a colleague of the operator.

He was a man, aged thirty-three years. Had been previously operated on twice by Bruns in Tübingen—first in 1888 for a tumor of the ninth rib, the rib being resected. Four mouths later a recurrence called for resection of the eighth rib. When seen by Hahn, in 1888, the growth had attained the size of two fists, springing from the lower ribs. The patient urgently requested the operation. Hahn resected the sixth, seventh, eighth, ninth, and tenth ribs. When the tumor was drawn out the pleura and peritoneum were extensively opened, and a piece of the peritoneum the size of the palm of the hand was excised, being involved in the disease. The diaphragm was wounded and secured to the upper edge of the external wound. The condition the next day was relatively good, but he very soon succumbed to the operation.

Case XXVII.

Alsberg's first case, 1888, chondroma (Riesenfeld, in *Deutsch. med. Woch.*, 1889, p. 321).

This patient was a female, aged twenty-two years. Operated on, July 15, 1888. There was a tumor of slow growth ; no pain ; no marked impairment of health. It must have been first noticed in April, 1888, when it was of a walnut size. When operated on in July it was a tumor of the size of a goose-egg, situated on the left side at the level of the seventh and eighth ribs, between the anterior and middle axillary lines. The skin was freely movable over the tumor. The tumor had a length of 9 cm. and a breadth of 6 cm. The eighth rib, which was surrounded by the growth, was resected and a piece of adherent pleura, two inches in size, cut out. The lung collapsed, but the patient quickly recovered and the pulse remained good throughout the operation. The wound was closed by deep and superficial sutures, with iodoform gauze drain at the lower end, which was later substituted by a rubber tube on the fourth day when the dressing was changed. The evening of the operation the patient complained of dyspnœa, which increased steadily, and was followed by a cough on the second day. On the evening of the third day temperature was 38.6°, pulse 140 ; but under treatment this was rapidly improved, and from the sixth day the pulse, temperature, and respiration were normal. Drainage tube was removed on the tenth day. In four weeks the patient left the hospital, being completely well and lung showing normal expansion. Microscopically it was an osteochondroma with myxomatous degeneration. According to Caro (*Deutsch. med. Woch.*, January, 1893, p. 57), there was no return up to January 19, 1890. Quénu and Longuet say none after five years.

CASE XXVIII.

Roswell Park's case, 1888, sarcoma, spindle and round-celled (*Annals of Surgery*, 1888, viii. pp. 254-257).

F. C., aged thirty-three years, of Machias, N.Y. Patient's family history good. Twenty years ago noticed a nodule, no larger than a pea, about the middle of the outer aspect of the left leg, where he had previously bruised it. Enlarged slowly until five years ago, when it began growing rapidly. It was removed and diagnosticated sarcoma by the doctor. It returned at its old site. Early in February of 1887 he consulted me. I found a hard mass the size of a fist embedded in the tissues on the outside of the left leg, a little above the middle. Skin not much discolored nor very adherent. Larger now than when first removed. I advised amputation. Dr. C. King, of Machias (who removed the first sarcoma), made it at the knee-joint; wound healed readily, with excellent resulting stump. A piece of growth sent me for examination proved to be a small, round-celled sarcoma.

By Dr. King's advice the patient returned to me in January, 1888. He presented a growth the size of a hen's egg a little above and to the outer side of the left nipple. It was fast, tender, but with movable overlying skin, apparently involving the whole thickness of the thoracic wall. Neighboring glands not involved. General condition good. Chest expansion normal, and on careful auscultation no difference detected between the sounds of the two lungs. No dulness on percussion in the neighborhood of the tumor. Main complaint, severe pain; also acknowledged he was losing a little flesh. I advised removal January 21, 1888, in my clinic at Buffalo General Hospital. Operation was made. Ether he took kindly. Skin over the tumor separated easily after a crucial incision had been made. It at once appeared that two, if not three, ribs were involved, and that total excision would be necessary. I began to separate periosteum on inner side of last rib involved, at short distance from edge of mass. Rib proved fragile, broke, and a spicule of bone forced through the pleura. As soon as the pleural cavity was opened I rapidly dilated the opening with my finger and determined that the growth was larger on the inner side of the thorax than on the outer; also, that there was adhesion in at least one place to the lung beneath. Having gone so far, I decided to extirpate the entire mass. I excised the tumor with the four ribs involved (fourth, fifth, sixth, and seventh), thus taking out a portion of the thoracic wall, some five inches in length by three and a half inches in width. After removing all the thoracic attachments I found that the band of adhesion connecting it with the lower border of the upper lobe of the lung was long enough to tie, and after throwing around it a strong ligature the mass was easily detached. During and after the removal a beautiful demonstration of the action of the heart in its pericardial sac was afforded. Hasty examination of the left lung, both ocular and by palpation, revealed numerous nodules scattered through the lung tissue of both lobes and on their surface. Had there been a single sarcomatous mass accessible I should have excised a portion of the lung. Under these circumstances such a measure was out of the question. Respiration but slightly disturbed. Pulse, however, became weak. Stimulants frequently given hypodermatically. I checked quickly what little hemorrhage there was and closed the wound with numerous continuous sutures. Over this iodoform was dusted and an antiseptic compress snugly bandaged down. At the close of the operation the face was slightly cyanosed, pulse 140, respiration 30 to the minute. Within an hour he was conscious, complained of great pain, checked by morphine subcutaneously given. During the ensuing night he was restless, requiring anodynes in large amounts. Pulse at one time 170, and very feeble. Following day temperature 97°, pulse 130, comparatively comfortable and taking sufficient nourishment. Next few days progressed very favorably, only once temperature as high as 101.1°, respiration fluctuated between 30 and 50 for two or three days. His left lung seemed to be inflated. January 27th, his face was cyanosed and anxious, respiratory murmur on left side lost; the vesicular character, which for three days it had had, became faint. Dressing had not been changed at all. Toward evening became delirious and died early the following morning. Five hours later I reopened the wound, which I found completely united by first intention. Cavity of the left pleura filled with a bloody serum, in which the lung seemed to have macerated, since it was soft and tore easily. Fluid had no odor. Hand passed into right pleural cavity; found right lung just as much studded with sarcomatous nodules as left.

Comments : Equal affections of the two lungs will account for the fact that on auscultation no difference was detected. Microscopical examination showed it to be a small spindle-celled sarcoma, while the nodules found in the lungs showed distinct sarcomatous elements, but of round-cells. It is of pathological interest to note how secondary growths differ from the parent tumor.

Case XXIX.

Tietze's third case, 1889, sarcoma (Reported from Fischer's surgical clinic in Breslau, *Deutsch. Zeitsch. f. Chir.*, 1891, Band xxxii. p. 424).

A. G., aged twenty-eight years, wife of a harness-maker. Family history good. Menstruated at sixteen. At the age of nineteen, while dancing, received a blow on the left breast, and had pain for some days. A hard lump developed on the outer side of the left mamma, about the size of a walnut. At twenty-two she married, the lump being present, but causing no inconvenience. During her first pregnancy, in the second year of her married life, the lump became larger. She was twice pregnant afterward. The tumor became larger during lactation. During last pregnancy (the fourth), the tumor attained the size of a man's fist, and increased rapidly in size during lactation with marked pain, which necessitated weaning.

She presented herself at the surgical clinic August 12, 1889.

Examination : No enlarged axillary glands. Dimensions of tumor : length, 11 cm.; breadth, 15 cm.; height, 12 cm.

Operation : Circular incision extending toward axilla, exposure of capsule, which was easily separated from the flaps. Tumor seemed to originate from the chest wall, extending toward the sternum and thence to the cartilage of the fifth rib. The adjacent ribs were separated so that the tumor seemed to be surrounded by a bony ring. The fifth rib being cut and the tumor being lifted, there were found both pericardial and pleural adhesions. The pericardial adhesions offered no difficulty, but in separating the pleura from the tumor this membrane was torn in several places, and during the manipulation the severed rib broke, permitting the entire removal of the tumor. The larger portion of the pericardium was now to be seen. When the pleura was torn the lung collapsed. The pleural rents were sewed up, but the sutures partly cut out, giving rise to some emphysema, which was controlled by a rather stout bandage. There was no material change of pulse while the pleura was open. The blood loss was small. The general condition of the patient after the operation was good, the pulse being moderate and strong.

For the first two days the dyspnœa was noticeable. The temperature on the evening of the operation was 37.8°, second evening 38.4°, the highest point reached. Sixth day, change of bandage, removal of iodoform gauze and sutures, the wound being healed primarily. The patient was dismissed on the fourteenth day, after the second change of bandage.

She did not report again until January, 1890. She seemed in perfect health. The thoracic defect measured from the fourth to the sixth ribs 5 cm. in length by 10 cm. in breadth. The heart pulsation could be plainly seen and the apex readily palpated with the fingers. The lung seemed to be functioning normally. Her household duties did not permit her remaining long enough for more extended examination. Up to the time of this report (1891) the woman had not reported since 1890. Quénu and Longuet say no recurrence after three years, but their reference is the same as mine. I think it should be one year, as she had not reported since 1890.

Case XXX.

First case of Vautrin, 1889, sarcoma (*Huitième Congrès de Chir.*, Lyon, 1894).

Mlle. X., of Thiancourt, aged nineteen years, entered the hospital September, 1889, to be operated on for a voluminous tumor of the breast, which had been observed only six months.

Of sound parents and in perfect health until the last few months, she is now pale, anœmic, her appearance one of suffering. All these altered traits, combined with her red hair, indicated a nature delicate and without resistance.

The right breast was infiltrated by an enormous neoplastic

mass, extending from the sternal line to the axillary border of the scapula, and from the clavicle to the last ribs. Skin wrinkled, veins dilated, distended, and glistening; the nipple umbilicated. She complained of great pain in the right arm. Impossible to mobilize the tumor or its attachments. Auscultation and percussion denoted no intrathoracic invasion.

His diagnosis was encephaloid sarcoma of the breast of rapid progress, with adhesion to ribs and also spreading in the intercostal spaces.

The operation was done September 17, 1889. Vautrin made an elliptical incision around the principal part of the tumor, whose limits were uncertain He dissected away the skin and laid bare the deep layers, more or less invaded, and arrived at the neoplasm, which absorbed all the mammary gland and pushed out feelers even into the muscles toward the axilla and under the pectorals. He commenced the dissection at the base of the tumor toward the sternum, where there were solid adherences with the ribs. He cut with a bistoury the cartilages of five successive ribs, then exposed the pleura on their internal surface; but at 7 cm. from the sternum he arrived at a place where this membrane was absorbed by the tumor. Here two rarefied ribs broke. Light tractions on the tumor tore the pleura, and air entered the chest. Several successive syncopes followed, and the chloroform was stopped and two injections of ether given. After five minutes pulse was improved, respiration became more frequent, but regular, and the operation was continued.

The pleura was opened, pneumothorax was established, and the lung was collapsed against the vertebral groove. A piece of diseased pleura 8 cm. wide by 6 cm. in length was removed. The fourth, fifth, sixth, and seventh ribs were cut back of the vertical axillary line at the angles of the ribs. The ablation of the inferior fibres of the great pectoral muscle and a part of the minor was done, and as the third rib was already adherent and diseased it was resected to the extent of a five-cent piece. The arm-pit was cleaned to the axillary vein and quite a large neoplastic node removed. The external digitations of the serratus magnus and the external border of the latissimus were found absorbed by the neoplasm, and Vautrin abraded them. The wound was enormous, measuring 25 cm. in its transverse diameter, and extending vertically from a point situated 3 cm. from the clavicle to the eighth rib. In the midst was the pleural opening, through which could be seen the retracted lung. The toilet of the pleural cul-de-sac was made by wiping with sterile compresses, without irrigation. It was useless to try to close the pleural opening by approximation, the loss of substance being too great. Passed a drain from the bottom of the pleural cavity toward the posterior costo-diaphragmatic angle across the thoracic wall, behind and internally to the angles of the ribs. The difficulty now was to close this immense wound. The flaps of skin and subcutaneous tissue, carefully husbanded during the operation, served partly, but had to be supplemented by a flap taken from behind the wound and from the axilla, and with a pedicle large enough to insure its vitality. The wound was united with silkworm-gut, leaving an opening 5 cm. in diameter. The operation lasted one hour and a half.

For two days she had attacks of vomiting and dyspnœa. The pulse remained regular at 100. The temperature rose the second day to 38.5°, and oscillated between 38° and 39° for about a week. About the third day the right lung commenced to respire in its whole extent, and by the fifth its rhythm was normal. The drain discharged at first a suspicious serum, then frank pus.

It is to be noted that the patient complained during eight or ten days of an intense pleuritic stitch, but at no time was the situation grave. About the fifteenth day she could sit up, and on the twenty-fifth the posterior drain was removed. The cicatrization was complete about the fortieth day, when the photographs were taken. It was observed that the lung was intimately adherent to the whole extent of the cicatrix, the sucking-in of which toward the cavity was easily seen during the respiratory movements.

The patient left the hospital about the fiftieth day, thor-

oughly well. Quénu and Longuet report no recurrence after
five years.

CASE XXXI.

Bardeleben's case, 1890, sarcoma (H. Plitt, *Inaug. Diss.*,
Berlin, 1890).

Wilhelm S., aged thirty-eight years, admitted into Charité,
Berlin, January 20, 1890.

Family history good. January, 1889, he fell on the ice
and struck heavily on the left side. A carpenter's rule in his
pocket produced a severe bruise at the lower part of the chest
on the left side. Three days afterward there was a swelling
extending from the sixth to the ninth ribs on the left side.
This increased in size, accompanied by severe pains. Six
weeks before admission he noticed several little lumps over this
region, extending up toward the shoulder. The patient is a
moderately strong man, somewhat meagre. He had suffered
considerably, and lost much rest at night in consequence.

Examination : Left side distinctly bulging below between
the mammary gland and the mid-axillary line. The skin was
practically uninvolved and quite movable. A row of tolerably hard tumors, arranged in oblique direction from below
and behind, forward and upward, was seen, the lowest of
these on the ninth rib in the midaxillary line, the highest
close to the mammary line at the sixth rib. The lowest of
them was of a walnut size, the highest bean size. It was
difficult to say whether the soft parts or the bony were
involved. The seventh and eighth intercostal spaces were
covered by the tumor. The upper edge of the seventh and
the lower edge of the ninth ribs were easily felt, as well as
the spaces above the seventh and below the ninth. The
measurement of the tumor on the eighth rib was 13 cm.
The distance from the xiphoid process to the point of the first
lumbar spine was three and one-half inches greater on this side.

Operation January 25, 1890. The first oblique row of
small nodules was taken out with a part of the latissimus
dorsi. The ninth rib was then resected, and then the seventh
and eighth. In resecting the seventh and eighth a piece of

the pleura had to be removed, being involved in the disease.
Then the soft parts of the sixth and ninth intercostal spaces
were cut through. The ribs were first cut through in front
and then behind. Through this wound the lung, adherent
to the diaphragm, could be seen moving up and down, as,
likewise, the heart and the diaphragm. The left kidney even
could be seen glistening through the diaphanous diaphragm.
Further inspection showed that more of the eighth rib was
diseased, so that one and one-half inches more were removed.
The operation lasted ninety minutes. Nothing alarming
occurred—only the rush of air and temporary fall of the
pulse. A large drain was put in and the wound sutured.
The dressing was changed on the fourth day, again on the
sixth, when a new drain was put in, and on the eleventh
the drain was removed. Since February 12th the patient
has been getting occasionally out of bed, and on February
22d he was dismissed at his request, the wound being
healed.

March 3d. The patient returned to have some glands in
the left axilla removed. These were as large as a goose-egg.
They were taken out, and healing promptly followed.

12th. In the supraclavicular space a new glandular swelling was discovered. This was removed, as well as three
small glands discovered lying on the subclavian artery.

22d. The patient was discharged.

Examination of the tumor showed it to be alveolar sarcoma
of the eighth rib. Horizontal measure, 13 cm.; vertical, 7 cm.
The eighth rib showed fracture. Five months later the patient
was again seen and was completely well.

Case XXXII.

Case of O. Witzel, 1890, sarcoma (*Centralbl. f. Chir.*,
1890, xvii. 523. "Ein Verfahren zur Beseitigung des aku-
ten nach Penetratio der Brustwand entstandenen Pneumothorax").

Male, aged thirty years, no further information given, except that the tumor was of the size of the fist and situated
over the tenth and eleventh ribs of the right side.

The operation was done by Witzel about Easter, 1890. Under morphine-chloroform narcosis a U-shaped flap was made, exposing the tumor. The tenth and eleventh ribs were divided and the pleura opened with the scissors after cutting through the intercostal musculature. It was necessary to remove the posterior remnants of the ribs as well as the transverse processes of the vertebræ corresponding to the resected ribs. The pleura being opened, cyanosis appeared with dyspnœa, at first of high degree, but improving in two or three minutes, whereupon the fluttering lung slightly expanded. The heart here as well as in another earlier case of Witzel's (*vide* Case XIX.) suffered out of proportion to the respiration. Witzel attributed it to the flexion of the large vessels, due to the sudden dislodgement of the heart. This, according to Witzel, occurs toward the side of operation when the pleura is freely opened, but toward the opposite side in closed pneumothorax.

In this case the pneumothorax was overcome by a simple method. The cavity was filled with warm boric solution, thus converting the pneumothorax into a hydrothorax. A male catheter attached to an irrigator was introduced into the upper anterior end of the outer wound so that it ran parallel with the wall of the thorax. The outer wound, particularly around the instrument, was now closed with deep and superficial sutures, so as to be air-tight and water-tight, excepting a small slit at the apex of the pleural cavity. Across this slit sutures were passed, ready to be drawn tight later on. The pleura was now filled from the irrigator through the catheter with warm boric solution at blood temperature.

When no more air, even on gentle shaking of the patient, issued from the slit, this slit was closed by tying the sutures. Slight and very transitory cyanosis followed the introduction of the solution. By lowering the irrigator the fluid at once began to flow from the catheter, and as it flowed the respiration and the heart's action improved. The fluid being drawn, the catheter was withdrawn in such a way as to prevent the entrance of air.

The results of the operation exceeded expectations. The patient being raised, at once breathed quietly and easily. Percussion and auscultation showed the lung to be expanding even where the resection had taken place. Only posteriorly was a slight friction heard, supposed to be due to blood clots.

The course of healing was uninterrupted. There was for a time a short fistula due to a ligature, and behind there was a slight dulness along the lower part of the thorax, indicating adhesion.

Second case of Witzel, referred to in this article.

This is evidently the same as Trendelenburg's (see Case XIX.), and is, therefore, thrown out. Quénu and Longuet give a second case of Witzel, but the only authority for it is Witzel's article; but a careful study of this paper in conjunction with the *Inaug. Dissertation* of Baldus, Bonn, 1887, to which Witzel refers for a fuller account of the case, shows conclusively that he speaks of Trendelenburg's case (Case XIX. in this list). This second case of Witzel (Case XIV., Class B, in Quénu and Longuet) is, therefore, thrown out. The case is not found in the list of Gerulanos nor in that of Paget, but, however, given as Trendelenburg's.

Case XXXIII.

Case of Marsh, 1890, chondroma (*British Medical Journal,* June 14, 1890).

F. W., aged twenty years, delicate looking woman, having a good family history, was admitted to Queen's Hospital, Birmingham, March 19, 1890. For many years has had double suppurative otitis media, a sequel of measles, but with this exception has had good health. Her work (brushmaker) involves a considerable use of the shoulder by the pectoral muscles.

Two years ago noticed accidentally a lump on her chest, just above the right breast. It felt hard, but not painful. It has slowly increased in size until a few months ago, has grown more rapidly since, and as it is becoming a deformity she wishes something done. Growth now measures about three inches in its long diameter, is irregularly oval in shape, hard, slightly lobulated, painless, situated over the sternal ends of the third and fourth ribs and firmly attached to one of them, the skin over it being normal and movable. Under an anæsthetic, growth was exposed by oblique incision; soft tissues easily separated from it, and the projecting part cut off with chisel and mallet. Proved to be an enchondroma; hard externally, but of a crumbly consistence underneath, attached to the third rib, a portion still being left wedged tightly in the third interspace, pressing the fourth rib downward. On removing this with a gouge it was seen that the growth expanded beneath the ribs into a mass of the size of one's fist. Owing to crumbling consistence and want of support, could only be removed piecemeal. The portion of the rib from which the growth originated was cut away, and so, then, was the capsule, which was so blended with the stretched pleura that it was necessary to excise the implicated piece, measuring three inches by three inches. There was now an opening of irregular shape in the chest-wall. Each bleeding point had been at once secured. The pale, salmon-colored lung, covered only by visceral pleura, could be distinctly seen. The exposed lung moved synchronously with each movement of inspiration and expiration, forward and backward, to and from the anterior chest-wall, rather than upward and downward, through a space of fully three inches. The amount of the movement varied with the length and the depth of the respiration, from one inch to contact with the chest-wall—in fact, great care had to be exercised to avoid including the lung in the sutures closing the wound. These physiological facts I leave to physiologists to explain.

It was impossible to bring the edges of the parietal pleura together, so I sutured the soft parts together over the openings, placing a short, good-sized rubber drainage-tube in the lower angle, anticipating there would be some pleuritic effusion. Patient fairly comfortable, though temperature kept at 100° F. for forty-eight hours. Dressings were now changed, and as they were dry and the percussion-note over the base was that of pneumothorax, I directed the patient to inspire deeply, to empty the pleural cavity as much as possible of air, and at the height of the inspiration I withdrew the tube and closed the wound with a strapping. Patient more comfortable after removing the tube; temperature fell to normal; wound healed by first intention. The slight pneumothorax disappeared. Patient shown at Midland Medical Society on April 23d. Percussion sounds everywhere normal and clear.

CASE XXXIV.

Alsberg's second case, 1891, sarcoma, round-celled and spindle-celled (From Caro, *Deutsche med. Woch.*, January, 1893, No. 3, p. 57).

Female, aged thirty-one years. She sprang from healthy parents and had never been sick before. In July she felt tearing pains in the right arm. These increased and kept her from her work. At the end of July she noticed in the right side above and anteriorly a tumor, which slowly but steadily grew. The pain forced her to apply to the hospital in Hamburg in the latter part of September. She was a woman above middle size, with large bones and well developed musculature. She had the appearance of suffering. The tumor, of a fist size, occupied the space between the clavicle and the fourth rib, and extended from the right parasternal line to the anterior axillary line. There was tolerably sharp limitation. The skin was red from repeated rubbing. The tumor, which was elastic, was firmly fixed to the ribs. There was no infiltration and no glandular involvement. Puncture brought only blood. The chest and upper organs healthy.

Operation on September 21, 1891. Alsberg operated. Made transverse incision through the skin and muscles. The tumor was found to lie directly on the pleura, in the second intercostal space. A second incision was made toward the axilla, and the base of the tumor separated from the pleura without opening it. The tumor ruptured during the manipulation. Seven cm. of the second rib was resected. In taking away the tissues of the second intercostal space a piece of the pleura about the size of a five-mark piece had to be cut out. The wound was immediately stuffed with iodoform gauze; the wound was finally closed by suture, a gauze drain having been put in. The narcosis was not disturbed by the temporary elimination of the right lung from the respiratory work. Toward evening the respirations were 48, pulse 90, temperature 38.7°. She had a bad night.

September 23d. Pulse was 120 and the patient seemed very ill; dressing was changed; hæmatothorax; six to eight tablespoonfuls of bloody serum drained out from the wound. For some days the bad symptoms continued.

28th. The wound looked well and the patient seemed better.

30th. She was out of bed, and on October 4th discharged. The tumor was a round-celled and spindle-celled sarcoma.

November 6, 1892. The patient died of inoperable recurrence after small operations had already been done in March, May, and June previously for recurrence beginning one month after first operation.

Case XXXV.

Case of Mikulicz, 1891, myxochondro-sarcoma cysticum (Reported by v. Noorden, *Deutsche med. Woch.*, 1893, xix, 346).

A man, aged fifty-four years. There was no hereditary history and no history of injury. Four years before a tumor of the size of a hen's egg was observed in the left anterior axillary line, about the seventh rib. This grew slowly for a long time and then more rapidly.

Examination, June 25, 1891. The tumor was pear-shaped; the circumference at the base was 78 cm.; vertical measurement 35 cm.; antero-posterior 26 cm. Reaches from the axilla to the bottom of the thorax, spreading forward under the nipple beyond the anterior axillary line, backward under the inferior axillary angle of the scapula to the external edge of the latissimus dorsi, and approaches posteriorly to within 1 cm. of the crest of the ilium. The shoulder blade was pushed slightly upward, but movable. The following regions were occupied by the tumor : the axilla of the left side, the lateral chest wall, the hypochondriac and mammary regions, and the region of the back lying beneath the scapula. The adduction of the upper extremity was diminished by the presence of the tumor. The skin over the tumor was movable, and there was nothing abnormal in the chest or abdomen. There was no sugar in the urine, no albumin, no indican, but abundant sedimentum lateritium. The hæmoglobin mounted up to 85 (v. Fleischl). Absolutely nothing else abnormal anywhere.

Diagnosis rested upon the historical data, the local and general condition, and upon the examination of a piece of the tumor excised for the purpose. (This, if enchondroma, would not invariably show the character of the tumor.) The situation, duration, the insignificant influence on the health, the absence of lymphatic glandular involvement, the extension of the tumor, the absence of skin infiltration implied chondroma, but always with the reservation that a malignant change, suspected from recent rapid growth, might have supervened to invalidate the diagnosis. A valuable wrinkle was furnished by the blood examination. The amount of hæmoglobin contents of 80 to 90, before the operation, corresponds, according to his frequently made observation, to no malignant tumor; and it has been established by many experiences in his clinic that the return of the hæmoglobin to the original amount meant that no recurrence had taken place, or, at least, none of malignant nature. The operation was done under morphine-chloroform narcosis, and lasted three-quarters of an hour. The patient was laid on the right side (opposite), and the operation was done in two stages. Incision was vertical. The involved diaphragm was cut off last, leaving an opening of antero-posterior measurement of 8 cm., being 9 cm. wide. The wound in the diaphragm was closed, partly by the muscles, especially the abdominal oblique muscles, and partly by direct suture. Some small tumors were found toward the vertebral column. The wound in the wall was enlarged, and the tumors removed with hammer, chisel, shears, and forceps. In this operation the seventh, eighth, ninth, and tenth ribs were resected. A large piece of the pleura was taken away, and a piece of the diaphragm, as before stated. There was no change in the pulse or respiration when the pleura was opened.

On the eighth day there were respiratory difficulties. Then percussion showed dulness up to the middle of the scapula. Punctures in the posterior axillary line brought one and one-half quarts of bloody fluid. Stitches were removed in fourteen days. He was discharged in five weeks, and went to work. Nearly a year afterward some small tumors were found near the lower end of the incision. These were subcutaneous, and were easily removed.

After fourteen months an examination showed sinking in of the thorax over the site of the defect; clear percussion to the ninth rib; vesicular breathing everywhere. Bulging out of the wall in the defect took place in expiration to the size of a fist. No recurrence is mentioned.

Case XXXVI.

König's third case, 1891, sarcoma (round-celled). (Paget's *Surgery of the Chest,* p. 180, case 21.)

This was operated on in 1891. A man, aged thirty-four years. Noted only six weeks; not painful; says he had a cough and was losing flesh. Looks thin, pale, and feeble. A firm, elastic growth, the size of a large apple, was situated over the seventh rib, in the axillary line. There was no fluctuation. Tumor was an alveolar round-celled sarcoma. The seventh rib was resected with adherent pleura; lung was found fixed by old adhesion of the upper lobe; there was no serious dyspnœa either during or after the operation. He recovered slowly from the operation, but there was rapid recurrence, which ended in death.

Case XXXVII.

Vautrin's second case, 1891, carcinoma (secondary to mammary carcinoma). (*Huitième Congrès de Chirurgie,* 1894, p. 162.)

This was a woman, aged forty-six years, housekeeper, of

excellent previous health, and the mother of several children. Vautrin had operated on her in October, 1890, for carcinoma of the right mammary gland. The axilla had been cleaned out and the aponeurosis of the great pectoral removed. Six months later there was recurrence, but she came to Vautrin only in December, 1891—that is, at least four months after the recurrence first manifested itself. The neoplasm was seated on the anterior aspect of the thorax, at the level of the old cicatrix, which was ulcerated in the region lying between the sternal and axillary lines. Adhesion between the tumor and the ribs was quite firm. There was no implication of the lung or bones by metastases, and there was no cachexia.

Operation: Circular incision about the tumor, excision of a considerable part of the great pectoral in the neighborhood ; resection of 7 cm. of the fifth and sixth ribs, beneath which the pleura was found adherent and diseased. The fourth and fifth ribs being also removed to the extent of 4 or 5 cm. and the hæmostasis being attended to, the pleura was opened, the chloroform being suspended. The air rushed in, the lung collapsed, continuing, however, after a fashion, to perform its functions; then the respiration became irregular, and the pulse weak, recovering itself intermittingly. Soon the heart and respiration became regular, and the pleural flap invaded by the disease was cut away. The pleural opening measured $3\frac{1}{2}$ cm. in diameter, and through it the healthy lung could be seen to fall back against the vertebral groove. It was impossible to unite the edges of the pleural wound, so he conceived the idea of attaching the parietal to the visceral pleura. The lung was seized with a pair of forceps, but, being caught too near the hilus, it could not be sufficiently mobilized to bring it into the opening. He made another attempt, catching the lung farther out, but again the required traction was too great. Finally, Vautrin enlarged the wound in the wall by a further resection sufficient to enable him to detach the pleura and sink it in until it could be sutured to the visceral pleura. The wound in the thoracic wall was thus filled in. He was much bothered by the resulting funnel-shaped wound, which the lung traction exaggerated in depth and which the absence of soft parts prevented him from filling in. The exposed surface was diminished as far as possible, but he was forced to leave uncovered the infundibulum formed by the visceral pleura at the bottom and the external surface of the parietal pleura all around.

On the third day, on removing the dressing, he was reassured. The lung had resumed its function, the pneumothorax had in a large part disappeared, and the air, filling out the alveoli of the lung, had caused the infundibulum, which had existed two days before, to become obliterated. The bottom of the wound gradually approached the level of the resected ribs.

The cicatrization was completed in two months, and the woman resumed her occupation. Four months after this— that is, six months after the intervention—a new recurrence was discovered in the skin near the sternum, but the woman declined to submit to any further operation.

Case XXXVIII.

Case of Zarubin, 1891, sarcoma (Jacobson's *Operations of Surgery*, 1897, p. 581).

Zarubin, of Kharkov, relates (*Transactions of the Kharkov University Society*, 1891, supplement to *British Medical Journal*, August 1, 1891) the case of a young Cossack, who sought his advice for a steadily growing and occasionally painful tumor of seven years' standing. It measured 21 cm. horizontally and 19 cm. vertically, occupying the right side of the chest between the nipple and the post-axillary line from the sixth to the ninth rib. The new growth was hard, nodulated, immovable, and non-adherent to the skin. The integuments over it were thinned, but otherwise normal, and the nearest lymphatic glands apparently unaffected. An osteochondroma of the thoracic wall was diagnosticated. The huge mass was removed, together with the involved portions of the seventh, eighth, and ninth ribs. The gap left in the chest measured

17 cm. in a horizontal and 16 cm. in a vertical direction. On
opening the thoracic cavity the lung collapsed, but only partially, owing to pleural adhesions around the periphery of the
new growths. No serious respiratory or cardiac disturbances
occurred, and the hemorrhage was trifling. The cavity was
gently cleansed with gauze, soaked in a 1 per cent. solution
of boric acid, and the skin wound, conical in shape, closed.
The growth was much larger than an adult's head, and weighed
six pounds. For the first two days the patient was much collapsed and cyanosed, and suffered from agonizing cough and
obstinate vomiting.

A long drainage-tube was inserted and the crucial-shaped
wound closed with silk sutures. Pus escaped from the tube
for eight weeks. On the sixty-first day the tube was removed
and the wound healed up. On the one hundredth day the
man was discharged with an appropriate pad over the deep
depression.

Case XXXIX.

König's fifth case, 1893, chondroma cysticum (From
Paget's *Surgery of the Chest*, Case No. 23, p. 180).

This was operated on in 1893. A man, fifty-five years
of age. Noted six or seven years before; gradual in its
development; for the last year rapid, and painful on coughing. The growth was firm and hard, but fluctuating over
its lower aspect. It was situated below the right nipple
over the fourth to the seventh ribs. It was a cystic chondroma. The fifth and sixth ribs were resected subperios-
teally. The pleura was opened. The lung collapsed. The
wound in the pleura was sutured save for a hole for drainage. There was intense shock. Death a few hours after the
operation. At the post-mortem examination half a pint of
blood was found in the cavity. There were chronic degen-
erative changes in the heart and the lungs.

Case XL.

König's sixth case, 1893, chondroma cysticum (From
Paget's *Surgery of the Chest*, p. 170).

A man, aged twenty-nine years, in good general health, had for ten years noted a growth on his right side in front of the shoulder. It had of late increased rapidly. Its lower part had become soft, and had been punctured and fluid had been drawn off. It now extended from the second rib down to the nipple; passed forward an inch beyond the middle line and backward to the posterior axillary line. It was nodular, hard as cartilage, fluctuating on its lower aspects, and almost as large as a child's head.

Operation April 26, 1893: A large flap was raised, including part of the pectoral muscle. The third rib was cut through and the growth was cleared with the finger and raised. It was found easier than had been expected to strip the pleura off it, except over its lower posterior aspect, where growth, pleura, and lung were all adherent together. At this point a piece of the pleura and a small piece of the lung were removed with scissors. Little blood was lost; pulse and breathing remained good. The day after the operation he had a good deal of dyspnœa; some suppuration occurred; he made a good recovery. The growth was cartilaginous, with cystic degeneration.

(This case is found in Paget's text, but is not included in his tables.)

CASE XLI.

Sheild's case, 1894, sarcoma of venter of the scapula and ribs (*Lancet*, London, 1894, p. 741).

Boy, aged ten years, sent to Mr. Sheild by Mr. Little.

One of seven children, three of whom are still living. Of the four dead, one died of tumor on the brain, one from "water on the brain," one was born dead, and one seemed to have been under term and lived only one week. The father is living and healthy; the mother is also a fairly healthy woman. No history of tuberculosis in any branch of the family.

About twelve or fifteen months ago his mother noticed a lump in the right arm-pit, and so took him to the Royal Free Hospital.

When seen by Mr. Sheild there was found in the axilla a

rounded, fixed mass the size of an orange, pushing up the shoulder and obviously compressing the axillary nerves. It was exceedingly difficult to make out whether or not the tumor moved with the scapula, but it appeared to do so slightly. The skin over the tumor was not inflamed and not adherent. There was no tenderness except on firm manipulation. The measurements of the tumor, taken with the arm abducted as far as it would go, were : From above downward two and one-half inches, and from before backward two inches. From in front there appeared to be marked wasting of the right pectoral muscles, and the deltoid was likewise much wasted, the head of the right humerus and the acromion being much more prominent. The right shoulder was held about one inch higher than the left, and the head was kept slightly bent to the right. The anterior edge of the trapezius was more prominent than the left. With the arms held to the side the tumor was slightly visible, but when abducted the tumor was very evident. From behind the right supra-spinatus and infraspinatus muscles were considerably atrophied, and the tumor was perfectly evident with the arm hanging by the side, forming quite a bulge along the axillary border of the scapula. There was well marked curvature of the spine in the dorsal region, the convexity being to the left, so that the inferior angle of the scapula stood out prominently. The right shoulder measured one inch more than the left in circumference, but the arm and forearm were more wasted than the left. The movements of the right shoulder were much restricted.

The gravity of the case was fully explained to the parents, who, nevertheless, desired that the operation be performed. Accordingly it was undertaken by Mr. Shield on August 13th. The A. C. E. mixture was used. An incision was first made into the axilla to define the limits of the tumor. It was now quite evident that the tumor was of scapular origin, but the finger could not pass between the growth and thoracic parietes toward the apex of the axilla. On moving the arm and scapula the tumor moved slightly also, and the opinion was formed—erroneously, as it proved—that if the

scapula was removed the tumor would come away with it,
leaving the thoracic parietes intact.   Accordingly two very
free incisions were made posteriorly—one from the superior
to the inferior angle of the scapula, the other almost at right
angles to this.   The muscles at the vertebral border were
divided ; next those at the superior border and angle; the
acromion was divided and the shoulder-joint freely opened.
The amount of blood lost was quite inconsiderable.   The axillary border was next attacked, and, as the bone could now be
separated from the chest, it became evident that the subscapularis and teres muscles no longer existed as such, but were
incorporated in the tumor mass, the growth, firm and fibrous,
entirely infiltrating their substance.   It was also ascertained
that the tumor projected markedly toward the thorax on the
upper aspect of the inner axillary wall, and the intercostals
and serratus magnus were implicated in the growth.   Seeing
the hopeless nature of the case if abandoned, an attempt was
made to peel away the intercostals and pleura from the tumor,
but this was quite impossible, and in attempting it the thoracic
cavity was opened, and the finger could plainly feel the growth
projecting upward against the ribs and lung, though, of course,
covered by the firmly adherent pleura.   Under such desperate
circumstances the parts were readjusted and the wound dressed
antiseptically.   Obvious pneumothorax existed, and the heart
became displaced, but there were no signs of shock.   The boy
died on the following day.

The autopsy showed the lung collapsed, with pleura non-
adherent.   The first and second ribs were firmly incorporated
with the tumor.   The growth appeared to have originated
in the subscapularis, and, though it apparently sprang from
the scapula, this was really not the case.   The whole subscapularis was practically a sarcoma, and the growth had in
like manner invaded the teres and serratus magnus and the
intercostals.   At the upper limit the main vessels and nerves
passed through its substance and were incorporated with it.
Mr. Targett, of the Royal College of Surgeons, of England,
further examined the tumor and found the bone full of it.

The microscope showed it to be a dense spindle-celled sarcoma, approaching to fibroma in some parts.

CASE XLII.

Case of J. L. Faure, 1895, sarcoma (*Revue de Chirurgie,* May 10, 1898, p. 402. Quénu and Longuet's article).

Man, aged sixty-one (in Quénu and Longuet's table, p. 399, the age is given as twenty-one), years old. Admitted to Bicêtre in 1895.

In 1885 he suffered a fracture of the rib at the same place. The tumor began eighteen months ago. Loss of flesh and progressive cachexia.

On the right side, at the level of the last ribs in the dorsolateral region, at 10 cm. distance from the vertebral column, was a tumor the size of the fist, elongated, about 12 cm. in diameter, hard, immobile, and difficult of limitation. The skin was movable; deep attachments evident.

Operation August 9, 1895, under chloroform. The tumor was reached as soon as the skin was incised. There were prolongations on all sides in the substance of the sacro-lumbalis and the antero-lateral muscles. It was impossible to think of raising it *en bloc*. It was extirpated *par morcellement*. While circumscribing a node as large as a walnut, projecting into the depths, the pleura was torn for about 2 cm. at the level of the costo-diaphragmatic cul-de-sac. The air went in and out of the wound. This rent was rapidly obliterated by a whip-stitch of catgut. The hemorrhage was considerable, blood flowing from all sides, perceptibly weakening the patient. Although some masses of tumor still remained, the wound was tamponed with iodoform gauze and sutured.

In spite of a deplorable general condition, after some days of extreme debility, the patient got better. The wound granulated and gradually contracted. The appetite returned after some time, and the patient got up.

November 10th. The wound is nearly healed, and the general condition relatively good, although it is evident that the tumor continues its evolution.

Case XLIII.

Karewski's case, 1895, sarcoma. *Soc. med.*, Berlin, January 29, 1896 (quoted by Quénu and Longuet.) *Deutsche med. Woch.*, 1896, No. 14 (quoted by Gerulanos.)

Quénu and Longuet tabulate this among operations of the first class in which the pleura was not opened. They remark that there were two extirpations, but only mention of the second operation that three ribs were resected. They give the date of operation as 1895, which evidently refers to the first operation, although no particulars of this are given. The patient was a man, but the age is not given.

This is undoubtedly the same case as that found in the table of Gerulanos, who classes it under the head of intrapleural or penetrating operations. He gives the date as 1896, and is evidently describing the second operation.

The patient was a man, aged thirty-six years, who had three walnut-sized tumors (sarcomas), situated in the axillary line of the right side. The sixth, seventh, and eighth ribs were resected with a piece of the pleura corresponding. When the pleura was opened there was great paleness, and the pulse became imperceptible. The man, however, recovered. Gerulanos says nothing of the final result, but Quénu and Longuet add a note to the effect that death took place with metastasis in the neck.

As I have been unable to consult the original (Karewski, *Deutsche med. Woch.*, 1896, No. 14), I have thus compared the data derived from these secondary sources and classed the case in the second group, the penetrating or intrapleural operations.

Case XLIV.

Quénu's case, 1895, chondroma (*Revue de Chirurgie*, May 10, 1898, xv. p. 396).

Man, aged forty-nine years. The tumor was situated on the left side and extended from the fifth to the ninth rib, and from the left parasternal line to the posterior axillary line. The tumor had grown slowly for ten years, then quite brusquely, with pains. In the operation the fourth, fifth, and sixth ribs were cut through, first in front of, then behind, the tumor, and the tumor removed with the ribs. There were two wounds of the pleura. There was at once cyanosis, apnœa, asphyxia, and fall of the pulse. The operation had to be finished very hurriedly. The respiration became regular again only after obturation of the wound. Death occurred on the fourth day of dyspnœa and collapse.

Quénu, in this case (*loc. cit.*, p. 382), tried the effect of artificial respiration, but without much effect, the conditions improving only after the suturing of the skin prevented any further access of air to the pleura. The patient never recovered from the collapse.

Case XLV.

Dennis's case, 1896, sarcoma (Park's *Surgery by American Authors*, 1896, vol. ii.; article by Dennis, p. 293).

"The writer has recently operated on a case of sarcoma which was situated upon the axillary aspect of the chest. The patient was fifteen years of age, and was struck over the site of the tumor by a base-ball. Three months later a well-defined, hard tumor was present, with firm adhesion to the rib. The tumor was removed, together with rib to which it was attached, and also the underlying pleura, which was also involved. The lung itself was not implicated in the growth."

The illustration (Fig. 128, p. 292, *loc. cit.*) shows a tumor lying obliquely across the right side of the chest, running apparently from the parasternal line near the junction of the ensiform cartilage with the second piece of the sternum, upward to the axillary line to a point above the nipple. Although it is stated in the above brief reference that the pleura was opened, nothing is said regarding the occurrence of pneumothorax, and no mention is made of the result, either immediate or remote. It is to be presumed, however, that the boy recovered, as the case seemed to be a perfectly simple one in a young and healthy boy. I have been unable to find any published account elsewhere.

CASE XLVI.

Bayer's case, 1896, ostrosarcoma (*Centralbl. j. Chir.*, 1897, No. 2).

The case was that of a boy, aged thirteen years, who had a sarcoma of the right thoracic wall. The tumor proceeded from the eighth rib, and was the size of a child's head. A musculo-cutaneous flap, with the convexity downward, was raised from the tumor. After separation of the insertion of the cartilage of the diseased rib the dissection of the tumor from the parietal pleura went on nicely for some space, until a place was reached where the pleura was adherent. Here it suddenly tore. The tear was tamponed, and the loosening of the pleura proceeded with further. There was another tear and collapse. The operation had to be discontinued, and rapid suture of a great tear in the pleura was done. The half of the tumor already dissected was dragged out and a tampon of iodoform gauze was put into the wound, which was left open. The patient recovered rapidly, and was well enough to be brought again into the operating-room three days later—on November 3d. At the second operation it was shown that the tumor reached inward and upward nearly to the vertebral column, and during the peeling away of this part of the tumor the pleura was widely torn again. He saw the lung completely collapse and sink back into the cavity. The patient seemed to sink. Bayer quickly drew the superior lobe of the lung up with the forceps out of the cavity of the thorax through the broad split in the pleura and fastened its under edge by means of stitches, separated about 3 cm. from each other to the periosteum of the sixth rib bounding the pleural tear above. At once the patient recovered himself, and the upper part of the lung was seen breathing regularly. The under lobe remained collapsed; so Bayer determined to leave the opening which remained after the previous operation, and which was still packed with the gauze put in at that time, in order to use it as a drainage opening and to avoid the gathering of secretions in the pleural sac. The course subsequently was quite smooth. There was some dyspnœa in the first few days, and the dressing had to be changed on account of soaking. He had, as a matter of precaution, not sewed the great wound, but, after packing the lower pleural opening, had merely laid iodoform gauze over the upper part and dropped the flap over that. The lung in its upper part—that is, behind the whole extent of the flap—was functioning beautifully, and had he been able to fix the lower lobe also there would probably have been accomplished a rapid expansion of that lobe also, and there would have been no dyspnœa at all. The suture caused no trouble in the lung itself. As he wrote, on November 30, 1896, the wound was completely healed, the flap being drawn in considerably into the defect. [Bayer thinks this plan of fixing the lung to the thoracic wall is certainly a very simple procedure, calculated to permit this important organ to resume its function quickly, and I can quite agree with him, as it was quite effective in my own case, and did absolutely no harm. I do not think it necessary, however, to attempt to fix every part of the lung. All that is required is to sew the lung into the opening in such a way as to obturate completely the pleural cavity. The whole lung will gradually expand.]

CASE XLVII.

Thompson's case (Reported by Davidson, 1896, sarcoma, *Texas Medical Journal*, Austin, 1896-1897, xii. 415).

Mrs. L. W., aged forty-five years, married, admitted to John Seely Hospital, Galveston, February 14, 1896. No history of any tumor-growth in family. She has had six children, the eldest being twenty, the youngest six. Patient's health has always been good. Gives no history of syphilis, gout, rheumatism, or any constitutional taint; habits have been good and she has led an active life.

One year ago last August a small subcutaneous nodule about the size of the end of one's finger was noticed two inches below and a little to the right of the left nipple. Felt more like cartilage than anything else. No dimpling of skin; no pain. It increased in size, becoming more and

7

more rapid as it grew older. In January, 1895, it was the size of a large pecan, and last summer (August), that of a hen's egg. From this time on the tumor grew fast, and became painful. The pain was of a deep, boring nature; there was occasionally a prickling sensation toward its sternal end. The pain was constant, growing more intense, so that for the last month the patient had been kept awake at night.

Fairly well nourished. No digestive disturbance since tumor appeared. No axillary enlargement. Has fallen off about ten pounds in the last few months. Lungs and heart normal. Examination of urine negative. Inspection reveals a tumor about the size of one's fist on the left side of the chest, having a nodular appearance. It occupies the sternal portion of the fourth, fifth, sixth, and seventh ribs, and their costal cartilages. The bulk of the tumor is below and to the inner side of the nipple. The mass is hard and firmly adherent to the ribs. In moving the tumor the ribs move with it, just as if they were part of it. No points of fluctuation. Indistinct outline at edges. The skin is not adherent to the growth. No tenderness complained of upon palpation.

Diagnosis : Tumor hard, growing from cartilage; chondrosarcoma.

Operation : Prof. J. E. Thompson did the operation, 9 A.M., February 17th. An incision was made through the skin, beginning at the lower border of the sternal articulation of the third rib and extended in a semi-lunar direction downward and outward to the upper border of the sixth rib ; thence outward and upward to a point to the left of the nipple. Immediately below the skin the tumor appeared, free from breast tissue. A few bloodvessels were clamped and the flap retracted, carrying with it the mammary gland. To the left of the tumor the third and fourth ribs were exposed and freed from their periosteum, and then cut directly through with bone forceps. These two ribs were again cut about three-quarters to one inch more to the median line, thus allowing a small piece of these ribs and the parietal pleura to be removed. Now the lung was exposed, but not yet collapsed. The index and middle fingers were then passed into the pleural cavity and used as directors, and with a stout pair of curved scissors the cartilaginous portion of the third rib was cut loose, then the sternal attachments of the fourth and fifth ribs, and, finally, the cartilage of the sixth rib severed. The pleura was included. Thus a hole about three inches in diameter was made; the lung was seen expanding under our eyes, the heart beating very slowly from the chloroform anæsthesia. The lung became more and more collapsed from exposure. Some bleeding from the intercostals and internal mammary. Bleeding stopped by pressure. The skin-flap was brought together with interrupted sutures of catgut. A plug of iodoform gauze was allowed to protrude through the outer part of the incision for drainage. Iodoform powder was dusted along the line of suture, a small strip of gauze and sterile cotton applied and held in place by a many-tailed bandage. Operation finished at 10.15 A.M. The pulse continued weak awhile after the operation, but the patient improved and gradually recovered, the wound healing by first intention. The gauze had been removed on the third day.

The tumor on examination was firm, cutting with resistance. A microscopical examination shows the structure of large spindle cells arranged in alveoli with considerable cartilage in the hyaline matrix; the giant cells scattered here and there in groups.

Contrary to expectation, the growth recurred. The patient returned to the hospital in July for further treatment. The neoplasm had recurred practically all around the line of the previous incision, involving the axillary space from the fourth to the sixth ribs. Prof. William Keiller operated. The patient left two weeks afterward. There was some tenderness and thickening over the third rib. The growth had returned and was growing rapidly, but was not associated as yet with much pain.

CASE XLVIII.

Doyen's case, sarcoma, year not stated (Quénu and Longuet's article, *Revue de Chirurgie*, May 10, 1898, p. 399).

In the table of cases it is stated only that the patient was a child and that the thorax was resected with adjacent pleura, giving rise to dyspnœic accidents. But the following is translated from Quénu and Longuet's paper, p. 383:

" We have observed," says Doyen, " one of these sudden syncopes upon opening the pleura, in the case of a young girl from whom we were extracting a wide sarcoma of the thoracic wall, which adhered to the ribs and reached to the parietal pleura. The diagnosis of a malignant tumor involving the pleura having been previously made, we spared intentionally, in dissecting out the tumor, enough flap to close the pleural orifice. The child suffocated as soon as the air entered the pleural cavity. The lung totally collapsed; a few efforts at coughing uselessly distended it, at the expense of the air still contained in the opposite lung. Another expiration, and death was imminent. With four strokes of the scissors a quadrilateral piece of the thoracic wall was removed, the skin flap was pulled over the gaping pleural opening, and the lips of the incision drawn together with forceps, aided by elastic pressure. Thereupon respiratory efforts became efficacious again. A few moments later the lips, up to this time pale and purple, resumed a pinkish tint, and the pulse became satisfactory. . . . In a few hours the little patient had recovered from this really thrilling operation. We have likewise observed grave dyspnœic symptoms in several operations on pulmonary cavities or hydatid cysts of the lung and of the pleura."

CASE XLIX.

Helferich's last case, 1897, sarcoma (Reported by Gerulanos in *Deutsch. Zeitsch. f. Chir.*, October, 1898, xlix. p. 498).

W. W., aged fifteen years, laborer, from Schlemmin, was admitted into Helferich's clinic April 6, 1897. Parents living and healthy. One brother died of diphtheria. No hereditary history. As a child had measles and diphtheria, otherwise has been healthy. In the summer of 1896 he received a heavy blow on his right side, in the region of the breast. With the exception of transient pains he had little trouble from this, and was able to go on with his work. After some weeks, however, he began to have stitch-pains on the same spot, which gradually increased in intensity, so that he was forced soon to give up his work and apply to the clinic for relief.

Condition on admission : Patient is of small stature, gracefully formed, of weak musculature and anæmic mucous membranes; otherwise not of sickly appearance. There exists disposition to cough; trace of mucous expectoration; no dyspnœa. Respirations 22 per minute. No pains on breathing, but some while at work. Thorax is weakly built. The supraclavicular and infraclavicular fossæ appear on the two sides alike, but there exists on the right of the mammillary line up to the posterior axillary line, from the third to the seventh rib, a flattened swelling, which is covered by unaltered skin. At this place the ribs cannot be felt. There is a slight left skoliosis. There is diminished respiratory movement on the right. Percussion is painful over the swelling. There is a dull area over the region of the swelling, with difficulty differentiated from the liver dulness. Over the swelling there is heaving respiration and pectoral fremitus. There are no heart disturbances, and the pulse is regular and about 90 to 100 per minute. Abdominal organs normal, urine shows nothing abnormal.

The differential diagnosis was difficult between encapsulated blood extravasation in the pleura, chronic disease of the ribs, and a tumor of the chest-wall.

The following operation was carried out on April 7, 1897 : An incision was made along the sixth rib corresponding to the prominence of the swelling. At the outer end of the wound the rib is almost substituted by easily bleeding granulations. This rib was cut off through sound bone. Toward the middle line the rib cannot be found, but instead only broken-down tumor masses. More deeply is a fist-sized cavity, filled with a pulpy mass of broken-down tissue mixed with blood. The bleeding continuing, the hole was tamponed and a bandage applied. The microscopical examination could not make out sarcoma, but rather pointed toward a caries of the rib. The subsequent course was quite smooth, and seemed to justify the expectation of ultimate recovery. The boy's general condition improved, and he gained steadily in weight. Only toward the end of May was there any change for the worse.

June 8th. Patient looks pale, almost cachectic. Temperature normal, pulse weak, 70 per minute. The cicatrix shows in several places small tumor masses, and is œdematous. In the middle of this cicatrix is an opening about the size of a thaler piece, surrounded by shiny, broken-down granulations. This opening leads into a cavity about 12 cm. deep, filled with soft masses. The whole region of the cicatrix is distinctly bulging. This bulging reaches from the parasternal line in front to the scapula behind, and from the fourth rib above to the eighth below. Percussion shows dulness from the mammary line at the fourth rib obliquely upward to the spine of the scapula, posteriorly; this area merges below with the liver dulness, toward the median side above reaches the parasternal line, and below corresponds with the heart dulness. Behind it reaches to the inner edge of the scapula below. The auscultatory signs are changed materially only in the region of the bulging.

There seemed now no doubt as to the presence of a sarcoma of the thoracic wall, and so operation was determined on.

This was carried out by Prof. Helferich on June 9, 1897, under chloroform. The old cicatrix was first surrounded by an elliptical incision carried through the sound tissues. From each end of this ellipse an incision was made, one forward to the cartilaginous part of the rib (sixth), the other backward to the scapula. The length of the incision was about 25 cm. A second incision was made parallel to the right edge of the sternum, about two finger-breadths from it. This was 10 cm. in length, and reached from the third to the seventh ribs. The flaps were dissected, one upward, the other downward. The inferior inner angle of this immense wound showed the seventh rib, which was resected to the extent of about 8 cm. The pleural cavity was opened here about three finger-breadths below and internally to the former cicatrix. The visceral and parietal pleuræ were not adherent. The finger introduced into the cavity came directly into contact with the dome of the diaphragm below and with the mediastinal wall internally. The tumor was found to fill the chest cavity at this point, so that apparently only a small quantity of lung tissue remained. The attempt was now made to lay bare the upper part of the tumor and to explore it. With this object in view the third rib was resected from its cartilage to the scapula, about 12 cm. in length, and the pleura was again opened. The finger could now feel that the upper surface of the tumor was, like the lower, smooth and free from adhesions. The tumor appeared likewise here to be non-adherent to the mediastinum. It reached from the sternum to the vertebral column, and seemed attached to the chest-wall over a large area. The upper lobe of the lung was free, but the middle and lower could not be felt. The whole space seemed occupied by the tumor. The space behind the cartilages was free, so that the finger introduced from above could be brought into contact with the one carried through the lower opening. A piece of gauze was now passed through this canal, and upon it the insertions of the fourth, fifth, and sixth ribs were cut through with the bone shears, together with the overlying soft parts. The internal mammary artery was not injured, but four intercostal arteries had to be first compressed and then treated by suture ligature.

Now the chest-wall from the third to the seventh ribs could be drawn outward from the sternum, so that the intrapleural relations could be more accurately examined. The tumor, which was of the size of a child's head, was firmly adherent to the lateral wall of the chest, and had incorporated with

itself the whole middle and inferior lobes of the lung; the
upper lobe was free and not adherent to the tumor.

Since the tumor had not advanced toward the mediastinum,
indeed seemed to be separated from it by normal lung tissue,
and since, between the tumor and the dome of the diaphragm
only loose adhesions were found, it was thought that a radical
extirpation of the tumor was possible, provided that a separation of the lung at its root could be accomplished. After
partly sharp, partly blunt, separation of the bronchus and of
the vessels of the upper lobe, which was done without great
difficulty, two large, strong forceps were put on the root of
the lung. No unusual disturbances of the heart or respiration were observed ; so the strength of the patient being well
maintained, the operation was proceeded with. The root of
the lung was now ligatured with a catgut ligature passed
through by means of a round needle. The lung tissue, also,
in so far as it was connected with the upper lobe, was ligated
with the assistance of a threaded needle and cut off. There
was no bleeding. The ligation of the bloodvessels and bronchi
of the severed root offered no special difficulty. A part of
the lung tissue which remained hanging to the stump was
so utilized that the root stump was completely covered over
with the serous layer. The tumor being freed from the mediastinum, and the adhesions with the upper lobe and with the
diaphragm separated, the mass was now connected only with
the chest-wall. The attachments of the tumor had required
the removal of the thoracic wall from the level of the third
rib down to the insertions of the diaphragm and back as far
as the vertebral column. For the separation of this vast mass
Helferich used the bow-saw with great advantage, doing the
work expeditiously and with very little blood loss. The route
of the saw was as follows : Above, through the third rib, the
tissues of the third intercostal space, the musculature of the
back, and the scapula horizontally under the spine; below,
the saw-line extended from the insertion of the seventh rib in
front to the junction of the eleventh rib with the vertebral
column behind, thus cutting through obliquely the seventh,

eighth, ninth, and tenth ribs, with the corresponding inter-costal spaces. The bleeding was easily controlled by compression and suture ligature of the severed intercostal arteries. The tumor was now attached only to the lung and to the vertebral column behind. The tumor being lifted and forcibly bent backward, the ribs were broken through at their necks, and the separation was completed with the bone shears and scissors. There now remained only a small part of the chest-wall attached to the completely collapsed upper lobe of the lung above and the abdominal part of the rib-wall below. The right half of the diaphragm and the right wall of the mediastinum were now in full view. The skin had retracted so far and the surface was so uneven that it was impossible to cover over the vast wound. It was therefore packed with warm gauze and a dressing applied.

The operation had lasted one and one-half hours. Neither when the pleural cavity was opened nor when the root of the lung was compressed were there any threatening signs, but the duration of the operation, the severe operative attack, the not inconsiderable blood-loss, and the marked chilling of the surface had brought about a degree of general collapse that required prompt attention. The face and mucous membranes were pale and somewhat livid, the pulse scarcely perceptible, accelerated, but regular, breathing superficial and rapid. Camphor injections subcutaneously and a litre of salt solution were given and warm packs applied. A few hours later the condition had much improved, the extremities being warm, the color much better, the pulse easily felt at 140, and consciousness had returned. Later he was still better, and the collapse seemed overcome, though he was still extremely weak. He took nourishment and expressed himself as feeling better. At six o'clock, June 11th, he seemed weaker, and the pulse could not be felt, the heart beating 160 per minute. Death followed gradually, and without special phenomena, at 7 A.M.

The autopsy showed marked anæmia of all organs. There was no infection of the wound and no sign of after-bleeding.

The left lung showed no change, and the heart was likewise found in a healthy condition. There was nothing abnormal in the abdomen, and nowhere was any metastasis to be discovered.

The tumor weighed 2130 grammes (about seventy-five ounces, or a little over four and one-half pounds). The measurements were: Height, 15 cm.; antero-posterior diameter, 16 cm.; distance from the outer aspect to the root of the lung, 15 cm. Seven ribs were found incorporated in the tumor, namely, the fourth, fifth, sixth, seventh, eighth, ninth, and tenth. In addition to these the third had been resected as one of the first steps of the operation, in order to get at the pleural cavity from above. The lengths of the various ribs resected were as follows: Third, 27 cm.; fourth, fifth, sixth, and seventh, 26–28 cm.; eighth, 19 cm.; ninth, 15 cm.; tenth, 9 cm. The fourth, fifth, and sixth were resected from their necks behind to their cartilages in front. The accompanying cuts give a very good idea of the size and relations of the tumor.

Case L.

Parham's first case, 1897; sarcoma.

L. B., aged twenty-seven years, colored, native of Louisiana; she has lived in New Orleans three years.

Family history good, personal history excellent. She was admitted to ward 36 of the Charity Hospital on February 16, 1897. She had noticed a swelling in the region of the right breast the previous summer. This grew rapidly. No further history could be obtained.

On admission a large, hard tumor, ovoid in shape, was found situated on the right side of the chest, apparently in the mammary gland. It was immovably fixed on the ribs, the whole chest moving when the tumor was grasped and forcibly manipulated. A diagnosis of osteosarcoma was made.

I did the operation February 22, 1897. The incision was a curvilinear cut transversely across the chest under the immense pendulous mammary gland. The muscles were cut transversely also. Nothing noteworthy occurred until the resection of the ribs, to free the tumor, was attempted. Several tears in the pleura now occurred while it was being separated from the growth, the pneumothorax being temporarily controlled by gauze tamponade. The pleura was rapidly separated from the tumor by pushing it away with the four fingers; but notwithstanding most careful manipulation, when the tumor was removed a rent in the pleura was found, five inches long, through which the whole hand could easily be thrust in. The air rushed in and collapsed the lung completely. It could be seen at the bottom of the cavity, against the vertebral column and mediastinum, making, it is true, feeble efforts to do its share of work, but it was practically eliminated from the labor of respiration. The condition for a time was extremely grave, and death on the table seemed imminent, the pulse failing to the point of almost total disappearance. While my assistants administered strychnine and digitalis hypodermatically, and attempted by artificial respiration to arouse the respiratory functions and the failing pulse, I tried to suture the rent in the pleura, but unsuccessfully, the stitches invariably tearing through when tied. The movements of the diaphragm and frequent coughing spells, during which the lung was violently blown out through the opening, made this an impossible undertaking. Realizing that this could not be done, I brought down the ragged pectoral muscle, placed it over the pleural rent, and sutured it there. When this was done her breathing improved, and she began to revive. The operation was completed by suturing most of the skin flap, leaving a large opening for drainage at the posterior aspect of the wound, which was packed with gauze. The mammary gland was not removed, as she had made a specific stipulation that it should be left. In the operation the third, fourth, and fifth ribs were resected, about five inches of each being taken away. The tumor sprang from the fourth rib, and involved the others by gradual growth and disintegration of their substance. The operation was begun at 11 A.M. and finished at about one o'clock. At the termination of the operation she was infused with salt solution. No record of the amount was kept, but it was sufficient to have a decided influence. The shock was very great, but her condition was improving, when returned to the ward, although she complained of nausea, vomited frequently, and was perspiring profusely. Strychnine, digitalis, and morphine were given at short intervals and continued at varying intervals for some days. When she was sent back to the ward her pulse was 132, respiration 62. At two o'clock that afternoon the pulse had fallen to 120, respiration 48. The next day the patient was still extremely restless, and had had a bad night. The pulse and respiration, however, continued improving during the second day, notwithstanding intense nausea and frequent vomiting. Temperature did not rise to 101° until the evening of the second day, when it reached 103.3° at 3.40 o'clock in the afternoon.

				Temp.	Pulse.	Respiration.
Third day.	Feb.	24.	8.30 A.M.	102.8	120	38
Fifth day.	Feb.	26.	8.30 A.M.	101.2	126	38
" "	"	26.	4.30 P.M.	102.7	130	46
Patient slept nearly all day.						
Sixth day.	Feb.	27.	....	102.8	....	...
" "	"	27.	4.00 P.M.	103.0	120	54
Seventh day.	Feb.	28.	9.00 A.M.	102.0	118	26
" "	"	28.	4.30 P.M.	102.9	....	...
Eighth day.	March	1.	8.30 A.M.	102.0	130	42
" "	"	1.	4.00 P.M.	102.0	116	52
Ninth day.	March	2.	8.30 A.M.	101.8	108	32
" "	"	2.	6.20 P.M.	102.0	128	52
Tenth day.	March	3.	8.30 A.M.	101.0	88	56
" "	"	3.	4 30 P.M.	102.0	104	54
Eleventh day.	March	4.	8.30 A.M.	101.4	106	32
" "	"	4.	4.30 P.M.	102 0	132	64
Patient had been restless during the first part of the night, but slept very well toward morning, without having had to take any morphine.						
Twelfth day.	March	5.	7.30 A.M.	101.0	101	68
" "	"	5.	4.30 P.M.	102.0	132	61
Patient has been sleeping nearly all day; does not seem to suffer.						
Thirteenth day.	March	6.	8.30 A.M.	101.5	108	52
" "	"	6.	4.00 P.M.	103.2	112	48
" "	"	6.	9.00 P.M.	102.2	108	50
Patient has been restless during day; not taking nourishment well.						
Fourteenth day.	March	7.	8.30 A.M.	100.7	120	48
Patient restless during night, slept very little; complained of nausea.						
Seventeenth day.	March 10.		9.00 P.M.	101.4	108	40
Nineteenth day.	March 12.		7.00 A.M.	101.2	108	32
" "	"	12.	9.00 P.M.	100.0	122	38
Patient coughing and quite restless.						

	Temp.	Pulse.	Respiration.
Twentieth day. March 13. 8.00 A.M.	99.8	112	36
Patient restless during night; coughing frequently.			
Twenty-first day. March 14. 8 30 A.M.	99.7	120	40
" " " 14. 3.30 P.M.	100.4	120	40
Patient did not complain much of pain during the night, and rested very well.			
Twenty-third day. March 16. 8.30 A.M.	100.0	108	30
Patient restless ; complained most of the night			
Twenty-fourth day. March 17. 8.30 A.M.	99.0	120	32
Patient slept and rested very well during night.			
Twenty-fifth day. March 18. 8.00 A.M.	99.0	100	28

From this time on the patient made steady improvement. She stayed in the hospital until April 19, 1897, when she was discharged in a satisfactory condition. The wound had suffered extensive sloughing of the skin and muscle, but on discharge was entirely healed. She was photographed May 8, 1897. March 20, 1898, the patient was again admitted to the hospital suffering from acute lobar pneumonia on the same side. She was discharged, well, just one year after operation—April 19, 1898. There was no recurrence whatever. (See Fig. 9.)

CASE LI.

Gauthier's case, 1898, sarcoma (*Revue Internationale de Médicine et de Chirurgie*, October 25, 1898, ix. No. 20).

Monsieur l'Abbe X., vicar, aged twenty-seven years, consulted me in September, 1897, complaining of violent pains on the left side and a little toward the back. Examination revealed nothing in the chest, and no tumor was apparent.

Diagnosis of intercostal neuralgia was made. Returned in October, 1897, again complaining and drawing attention to a slight enlargement situated on the left side, posteriorly, at the level of the ninth rib.

Operation was proposed, but declined. He returned in December, and then progress was evident. The tumor had attained the size of an orange. A fluctuating point justified Gauthier in announcing a diagnosis of cold abscess of slow evolution, developed on the diseased rib. He accepted advice, and was operated on, January 21, 1898. Notwithstanding this unusually slow evolution, Gauthier believed he had to deal with a cold abscess, and was disagreeably surprised by falling, with

the first stroke of the bistoury, into a mass having the microscopic characters of a cystic osteosarcoma. The hope of being
able to resect the pleura was soon shown to be illusory, for it was
intimately adherent. Gauthier had then to decide between an
operation, frankly incomplete, leaving half of the tumor in
place, or an operation, intrapleural, with operative pneumothorax, whose gravity he would not disguise from himself.

With the point of a bistoury a small opening was made a
little above the neoplasm, and this was immediately closed
with gauze, to permit the air to enter only very slowly and
to prevent the too brusque retreat of the lungs. After two
minutes he proceeded more rapidly with scissors to remove
a portion of the costal pleura as large as the hand. Having
tied the intercostal arteries, the wound was closed except
where a large piece of skin had been removed. This space
was packed with gauze. During the whole operation the
breathing was bad, and there were several bad attacks of
cyanosis. The chloroform, which was badly supported, was
stopped when the pleura was opened. Respiration became very
short and rapid, and it was necessary to give some injections of
ether and caffeine. There was, however, neither syncope nor
arrest of respiration. The patient having been carried to his
bed, he got an injection of one quart of artificial serum.

January 22*d.* Pulse 144, respiration 38, temperature 37°
to 37.8°.

23*d.* Pulse 132, 148, respiration 30, 35, temperature
38.2° to 39°.

Dressing was changed. Pleural adhesion already existing.
Pleural adhesion developed with excessive rapidity, so that,
the temperature being still up, an abscess was feared; but this
was not found until February 27th. About 400 or 500
grammes of pus was evacuated. The cavity was washed
and drained through rubber tubes.

March 15*th.* The tube escaped into the pleura. Gauthier
was called to get it out. In the month of May he was completely recovered.

Gauthier saw him in June. No sign of recurrence. Cicatrix was beautiful. At the end of August he returned with signs of recurrence. The ends of the resected ribs were enlarged and a tumor the size of a nut in the place of the old tumor. Operation was proposed, but he hesitated. He finally went to Paris to undergo operation under conditions infinitely more advantageous than the first, the lung being now completely surrounded by pleural adhesions.

The recurrence showed itself about six months after the operation and about three or four months after the complete cure by the first intervention.

Case LII.

Parham's second case, 1898, chondrosarcoma.[1]

J. R., aged twenty-eight years; ward No. 10, Charity Hospital. Admitted to service of Dr. Parham, July 24, 1898.

Previous History. Born in Iowa. Patient has led a roving life, but has lived in Louisiana for the past seven years.

Family History. Father died at sixty-eight years; cause unknown; mother living, as also three brothers and one sister.

Personal History. Measles when a child; malarial attacks twice in the last ten years; soft chancre ten years ago, followed by ulcers on the legs, scars of which can still be seen.

History of Present Disease. His occupation was that of a tinner. He frequently used a brace and drill in mending stoves, resting the brace against the left chest-wall. He never felt any pain, however.

He first noticed a swelling, about six years ago, about the size of an egg. The tumor grew very slowly, being last fall about the size of a turkey egg. Rate of growth has since this time been more rapid. Pain has been present during the past eight months. This pain, at first sharp and lancinating, afterward duller in character, seemed severer when in repose.

Condition on Admission. Was rather anæmic and pulse was weak, but he was otherwise in excellent condition.

Examination just before Operation. Heart, lungs, and urinary organs normal; general condition satisfactory.

---

[1] For the notes upon which this history is based I am indebted to Mr. B. Guthrie interne in charge of my service.

The tumor, which occupied the upper half of the left side of the chest, was firmly united with the ribs and extended from the clavicle to the sixth rib. It was a firm, solid tumor, ovoid in shape.

Operation, August 6, 1898, 9 A.M. The anæsthetic was chloroform, administered by Mr. Mason, interne. I was assisted in the operation by Dr. E. D. Martin, Dr. F. A. Larue, and Dr. J. D. Bloom, the house surgeon, who kindly attended to the management of the Fell-O'Dwyer apparatus.

Operation : A fan-shaped flap was outlined by an incision approximately parallel with the boundaries of the greater pectoral muscle, the muscle being cut through transversely below between the seventh and eighth ribs and along the sternum and inner half of clavicle. The skin and muscles were raised together. The minor pectoral was represented by only a few attenuated fibres, which could not be taken up as a distinct layer. The flap having been well turned up over the clavicle and covered with gauze, the tumor was now cleared thoroughly at its base and the ribs exposed. It was easily seen at this stage that the first rib was not involved in the tumor. The axillary structures were easily avoided by dragging them, with the pedicle of the flap, upward, toward, and over the clavicle and coracoid process. An attempt was now made to resect a portion of the fifth rib, in order to identify and begin the clearing of the pleura from the tumor, but this proved to be impracticable, the pleura being naturally so thin at this point. The cavity was unavoidably opened, and there was a sudden in-rush of air, which was, however, easily controlled by gauze packing. A second attempt was made to clear the fifth rib at another point. This was partially successful, but in the intercostal space above the pleura was again torn, and the patient showed sudden signs of collapse. The fifth rib was quickly cut in front and then in the axillary line behind the tumor. The lung was now almost completely collapsed, and there was a sudden fall in the pulse, almost to disappearance, the patient becoming blanched and showing all the signs of profound shock. The pleural openings were packed quickly with gauze, and Dr. Bloom was requested to begin the use of the Fell-O'Dwyer apparatus. As soon as the tube was inserted and the apparatus working the lung began to recover itself, and the man's condition at once improved. The respiration was now admirably maintained, so the operation was proceeded with. The fifth rib had now been loosened by cutting it through in front and behind. The soft parts between these two points were cut through quickly with the scissors, and then the fourth rib was cut through in the axillary line, the costal attachment being as yet unattacked ; then in turn the third and the second ribs were cut in the axillary line. Scissors were then run along rapidly through the first intercostal space, and the second, third, fourth, and fifth ribs were successively attacked in front, being cut through with the bone forceps. The remaining soft parts were cut through with the scissors, and the pleura, being peeled off from the tumor with the fingers, the growth was taken away. There was now a very large hiatus in the anterior wall of the chest, through which the lung could be beautifully seen alternately expanding and contracting. The assistance rendered by the Fell-O'Dwyer apparatus was evident to all. Whenever there was any hitch in working the apparatus the lung at once showed signs of collapsing, but when the apparatus was in working order respiration was almost as regular as the normal breathing. I can imagine no better demonstration of the usefulness of this admirable device. This artificial breathing was maintained until the lung could be drawn out sufficiently to be sutured with a continuous catgut suture to the margin of the opening. This being done, there was no further tendency to collapse of the lung, and so the Fell-O'Dwyer was discontinued. Before this suture had been begun, violent spells of coughing, which were rather frequent, somewhat delayed the operation, as with each attack the lung was forcibly extruded through the opening, the inferior lobe being driven completely out, so that it could be picked up in the hand. The lower lobe of the lung was never at any time pushed up by collapse so far as to expose the pericardium, so

8

that no distinct view of the pericardium and the working of the heart could be obtained.

It was observed by all present that the lung expanded in expiration and fell back in inspiration. The operation was completed by the uniting of the muscles and the skin by separate sutures. The duration of the anæsthesia was one hour and forty minutes. The man was sent to the ward in a very satisfactory condition. The wound was completely closed without drainage, the pectoral muscle being separately united.

August 6th. During the day the temperature rose once to 101.6°, pulse to 108. Respirations very little quickened.

7th. Day after the operation the temperature ranged from 100° to 100.8°, and the pulse only reached 100, being generally about 85. The respirations averaged about 24.

8th. Temperature never went up to 100°, pulse only once as high as 80. There was some slight dyspnœa, some pain in chest, and he was somewhat paler.

9th. Temperature up to 101°, pulse remaining at 80. Some dyspnœa; severe pains in the side.

Finding at my morning visit evidences of fluid in the chest, I returned in the afternoon and aspirated in the interscapulovertebral space 4 oz. of bloody serum and a considerable quantity of air. Removed all the cutaneous stitches and put on collodion and cotton. From the tenth to the fourteenth day the temperature kept above 100°, though the pulse was rarely over 85, the respirations ranging from 17 to 24, very regular and full. The left lung had evidently expanded considerably. As the dyspnœa was somewhat greater, the cyanosis more evident, and the physical signs showing accumulation in the chest, I again aspirated, drawing off ten ounces of sanguinolent fluid. More air came away than could be accounted for by leakage of the syringe. After this he breathed more easily, but the temperature rose by night, and on the next day, August 16th, reached 103° in the evening. The wound on its inner aspect had opened and was discharging pus.

18th. There is some slight subcutaneous emphysema in the axillary region. The temperature reached 103°. On the

20th it was at 2 A.M. 104°, but fell to 102° by 5.30 P.M., rising again at night to 103°.

22*d.* The temperature had been running high, reaching again 104° in the night. Opened up the wound at two places, finding two large abscesses, one antero-superiorly, the other at the lower and inner side. The abscesses apparently rested on lung substance. Washed out with hydrogen dioxide, followed by one per thousand formalin solution, and packed.

24*th.* During irrigation of abscess cavities by Dr. Larue there was a sudden and violent attack of dyspnœa, only arrested by turning the patient on the affected side.

25*th.* Doing better. The temperature has been constantly lower since the opening of the abscesses. For a long time the lower wound discharged, and a fistula could be traced backward for six or eight inches. The percussion not clearing up in this region behind, I contemplated a posterior resection for drainage, but gradually the discharge diminished, and, finally, on November 7th, it was found completely closed. I believe the suppuration to have been due to the silk stitch which I put in as a continuous suture, uniting the pectoral muscle. This stitch came away covered with pus.

December 3d. Came to my office for examination. The lung seems almost completely to have recovered its expansion power. The skin is drawn in on inspiration and is pushed out in expiration. On forced expiration or in coughing the lung herniates through the opening, producing a large protrusion of the skin. The hand can be pushed well into the defect, and the ventricle can be distinctly felt to the inner side of the defect. I believe some valuable investigations could be made into the physiology of the heart and its innervation, so accessible is it to palpation and the application of electrodes. Von Ziemssen has made a most elaborate study of the celebrated Fischer case (Case XI.), operated on in 1878, and also of one of Helferich's cases (Case XVII.), operated on later. I shall have a shield made to protect him from injury and to interfere with the development of hernia of the lung through the defect. A letter received from him the latter part of December, five months after the operation, says he is well.[1]

The tumor was a chondrosarcoma, starting from the third rib.

The accompanying photographs give a good idea of the size and relations of the tumor. The structure of the tumor was of hyaline cartilage. I am much indebted to Dr. O. L. Pothier, Hospital Pathologist, for the photographs and the photo-micrographs of this case.

Dimensions of tumor: Circumference, horizontal		.	. 3⁵ cm.
" " " sagittal	.	.	. 38 "
" " " vertical	.	.	. 35 "
Length of ribs removed, second	.	. . . .	. 10 "
" " " third	.	. . . .	. 13 "
" " " fourth	.	. . . .	. 12 "
Cartilage removed	. . . . .	. .	. 1 "
Weight	. . . .	. .	. 771 grammes.

The sternal cases of König and Küster are here reported more fully, as they illustrate some special points of interest. Other sternal cases are to be found further on in tabulated form.

CASE II.

König's case, 1882, sternal sarcoma (Leonhard Meyer, *Inaug. Diss.*, 1889).

This was a woman, aged thirty-six years. There was a tumor situated in the region of the sternum; it had been developing two and one-half years without much pain. It extended across the body of the sternum, without showing much laterally. It extended forward to the insertion of the second rib, being completely fixed to the sternum. The patient was otherwise healthy. König made a diagnosis of endosteal tumor and myeloid sarcoma of the sternum.

Operation, July 17, 1882. A transverse incision was made through the skin, the four flaps being dissected away from the sternum, exposing above, the manubrium, laterally, the cartilages of the ribs, and below the xiphoid process. The upper ribs were separated about 3 cm. from the sternal edge and freed of soft parts by the elevator. The sternum was sawed through at the level of the first rib with the bow-

---

[1] A note just received (May 26) assures me he is still well.

saw. Now followed the hardest part of the operation, namely, the lifting and loosening of the sternum and the tumor from the mediastinum. The sternum was held up and, with sharp hooks and the finger, cautiously pushed in between the tumor and the mediastinum. Both mammary arteries were exposed and tied double and cut. In spite of the greatest care the right pleural sac was torn at the level of the fourth rib. A prominence of the tumor was closely adherent to the pericardium, and had to be cut with scissors. The sac was injured and the right ventricle came into view. Then in the further dissection, the left pleura was torn. The openings in the pleura were quickly closed with gauze balls. The flaps were sewed from below upward in such a manner that the gauze balls could be removed. Drains placed in the lateral ends of the wound and a large dressing applied. The patient recovered well. There were no violent symptoms of any kind. On the fourth day there was great frequency of pulse, controlled by digitalis. The healing was *per primam*, except a small surface. [Patient died a year later of rheumatic fever, as I learn from Paget's account of the case, p. 173.]

CASE III.

Küster's case, 1882, gumma of sternum (*Berlin. klin. Woch.*, 1882, 127; also Baldus, *Inaug. Diss.*, 1887).

A man, aged thirty years, a picture of vigorous health, with no history and absolutely no signs of syphilitic trouble. He perceived, in 1882, a tumor on the right side of the sternum, with dull pains deep in the chest.

The tumor grew slowly. He was admitted to the Augusta Hospital, Berlin, October, 1882. He was put upon iodide of potassium, which was kept up six weeks without effect. A tumor the size of a goose-egg was observed on the right side of the sternum and closely attached to this bone at the level of the third and fourth ribs.

Operation October 27, 1882. The thorax being opened, the tumor was found attached to the right lung and pericardium. In loosening it from the lung a hole the size of a nickel was cut. The lung collapsed at once, as the gauze was not put in quickly enough. The pericardium was then freed without accident, and the tumor removed. A piece of gauze was put over the pleural rent and allowed to come out of the wound by the other end. Hand pressure was kept up over the wound, while the flap was pulled down and compressed by the other hand externally. The inner hand was then carefully slipped out while the pressure was maintained by the other hand externally. The wound was then closed up and sutured. The result was good. He was discharged in four weeks.

Küster discusses the question of diagnosis in these cases, and asks whether syphilitic tumor, when not yielding to iodide of potassium, should be removed. He answers this question in the affirmative. He refers also to König's case, published in 1882, who discusses the opening of both pleuræ and refers to his experiment on animals.

CONSIDERATION OF THE RESULTS.[1]

Immediate results or results of the operation.

	Series I.	Series II.
Died in consequence of the operation :		
Time not stated	0	1
Within twenty-four hours . . .	2	7
Within forty-eight hours . . .	1	1
During first week	1	3
During second week	0	2
During third week	1	0
During first month	1	0
After five weeks	1	2
Total deaths from operation .	7	16 = 23
Recovered from operation . .	19	35 = 54
Result not stated in one case .	1	... = 1
	27	51 = 78
Percentages of mortality of known results .	26	31.3 29.5

In many of these cases more than one operation were done on the same patient. In such cases the operations stated in the above table are those described in the detailed account of cases.

The following table exhibits the number of operations, the time patients lived after the first operation, after the last operation, and the final result as far as known :

[1] This table has been corrected to correspond with the transference of Case XVII. from Series II. to Series I.

NUMBER OF OPERATIONS DONE.

Series I.

No. of case.	Date.	Extra-pl.	Intra-pl.	Final results.	Incomplete op.	Time living after first op.	Living after last operation.
1	1678	1	......	Recovery.		?	
2	1820	1	......	"	?	2 years.	
3	1837	1	......	"	......	?	
4	1837	1	......	"	......	1 mo.	
5	1855	2	......	Death.	......	?	
6	1856	1	......	Death from recurrence.	......	6 years.	
7	1859	1	......	Death.	......	2 days.	
8	1863	1	......	"	1	?	
9	1866	1	......	Recovery.	......	9 mos.	
10	1869	1	......	Death.	1	24 hours.	
11	1871	1	......	"	1	?	
12	1871	1	......	"	1	1 mo.	
13	1878	1	1	"	?	9 mos.	5 months.
14	1882	1	......	"	1	3 weeks.	
15	1882	1	......	Recovery.	1	?	
16	1884	1	......	?	1	?	

If this table be corrected by the addition of the cases of Senn (Appendix, p. 138), Hamilton (Appendix, p. 139), and of Hartley and Steele (see appended slip), the statistics will read: .

Total deaths from operation in Series II. . 16

Recovered from operation 39

Total intrapleural cases . . . 55

Percentage of mortality in the intrapleural cases 29, instead of 31.3, as printed.

					tions.	operation	operation.
1	1818	?	1	....	......	?	
2	1855	......	1	....	......		
3	1861	......	1	Death.	......	2 days.	
4	1861	......	1	"	......	4 "	
5	1861	......	1 (?)	"	......	1st week.	
6	1861	......	1 (?)	"	1	13 days.	
7	1873	......	1	"	......	3 mos.	
8	1873	......	1	"	......	3 "	
9	1874	......	1	Recovery.	......	6 weeks.	
10	1878	......	1	Death.	......	?	
11	1878	1	1	Recovery.	......	2½ years.	1½ years.
12	1879	......	1	Death.	......	2 days.	
13	1880	......	1	"	......	5 "	
14	1880	......	1	"	......	1 day.	
15	1883–68	1	2	Recovery.	......	4 years. 10 mos.	1 yr. 4 mo. Inop. recur. Died 7 years after operation of pneumonia.
16	1884	(2)	1	"	......	....	2 years.
†17	1885	......	1	"	......	....	4 years.
18	1885	......	1	"	......	....	14 months.
19	1886	......	1	"	......	....	1 year. No recur.?
20	1885	......	1	"	......	3 weeks.	

* Figures in ( ) indicate operations of not sufficient moment to justify tabulation.
† This case has been by mistake included in this table; it should be in Series I.

Series II—*continued.*

No.	Date.	Extra-pl.	Intra-pl.	Final results.	Incomplete operations.	Time living after first operation	Living after last operation, with remarks.
21	1886–89	2	3	{ 4 recov } { 1 death }	......	3 years.	{ 14 days. Pleura { opened at 3d, 4th, { and 5th operations.
22	1887	......	1	Recovery.	......	4 mos.	Recurrence.
23	1887	......	2	Recovery after both.	. ....	3 years.	Recurrence after last operation.
24	1880	......	2	2 recov.	......	8 mos.	Not stated.
25	1888	......	2	2 "	......	4 yrs.and 8 mos.	2 years 4 months. No recurrence.
26	1888	2	1	Death.	......	24 hours.	
27	1889	......	1	Recovery.	......	5 years.	Q. and L.
28	1888	......	1	Death.	......	6 days.	
29	1889	......	1	Recovery.	......	3 years.	Q. and L.
30	1889	......	1	"	......	5 "	Q. and L
31	1890	......	1	"	......	5 mos.	Two operations for gland. recurrence.
32	1890	......	1	"	......	Not stat'd	
33	1890	......	1	"	......	Not stat'd	
34	1891	......	1	"	......	13 mos.	Small op. for recur. in soft parts.
35	1891	......	1	"	......	14 "	Small subcut. recur. removed.
36	1891	......	1	"	......	?	Death from rapid recur.; op. declined
37	1891	......	1	"	......	6 mos.	
38	1891	......	1	"	......	3 "	No further information.
39	1893	......	1	Death.	......	Few hrs.	
40	1893	......	1	Recovery.	......	Not stat'd	
41	1894	......	1	Death.	......	24 hours.	
42	1895	......	1	Recovery.	1	3 mos.	Recurrence.
43	1896	1	1	2 recov.	......	Not stat'd	Recurrence.
44	1895	......	1	Death.			
45	1896	......	1	Recovery.	......	Not stat'd	
46	1896	......	1	"	......	....	No report.
47	1896	1	1	"	1	5 mos.	No report.
48	1896	......	1	"	......	... .	No report.
49	1897	1	1	Death.	......	48 hours.	
50	1897	......	1	Recovery.	......	1 year.	No subsequent rep.
51	1898	......	1	"	......	6 mos.	Recurrence.
52	1898	......	1	"	......	5 "	
		9	58				

RECURRENCE.

Series I.

Sarcomas : Known to have recovered from operation—
    Nothing known as to recurrence 4
    Recurrence (Cases 15, 17, 22, 24) 4
    No recurrence after two years (Cases 2, 23) . . . . 2—10
Myxosarcochondroma : (Known to have recovered from operation—
    Recurrence . . . . 1—1
Chondroma : Known to have recovered from operation—
    Nothing known as to recurrence (Cases 9, 25, 26) . . 3
    Recurrence (Cases 3, 5, 6) . . . . 3— 6
Carcinoma : Known to have recovered from operation—
    Nothing known as to recurrence . . . . . . 1— 1
                                     18

Series II.

Sarcomas: Known to have recovered from operation—
Nothing known as to recurrence (Cases, 2, 22, 31, 32, 39, 45, 46, 48) 8
Recurrence (Cases 15, 21, 24, 34, 36, 42, 43, 47, 51) 9
No recurrence after 1 year (Case 50) ⎫
"   "   2 years ("   16) ⎪
"   "   3 "   ("   29) ⎬   .   . 5—22
"   "   4 "   (" *17) ⎪
"   "   5 "   (" 30) ⎭
Chondrosarcoma: Known to have recovered from operation—
No recurrence after 5 years (Case 25) ⎫
"   "   5 mos. (" 52) ⎭   .   .   .   .   . 2— 2
Myxochondrosarcoma: Known to have recovered from operation—
Recurrence (Case 23) 1
No recurrence after fourteen months .   .   .   .   . 1— 2
Chondroma: Known to have recovered from operation—
Nothing known as to recurrence (Cases 9, 33, 40) 3
Recurrence (Cases 1, 11) 2
No recurrence after 14 mos. (Case 18) ⎫
"   "   5 years (" 27) ⎭   .   .   .   . 2— 7
Myxochondroma: Known to have recovered from operation—
Nothing known as to recurrence (Case 20) .   .   .   . 1— 1
Carcinoma: Known to have recovered from operation—
Recurrence (Case 37) 1
No recurrence after one year (Case 19) .   .   .   . 1— 2

Total   .   . 36

SUMMARY OF RECURRENCES.

Series I.

	Recov-ered.	Result stated.	Recur-rence.	No recurrence up to				
				1 yr.	2 yrs.	3 yrs.	4 yrs.	5 yrs.
Sarcoma and its combinations, including chondrosarcoma . .	11	7	5	...	2			
Chondroma, simple . .	6	3	2	...	...	...	...	1†
Carcinoma . . .	1	0						

Series II.

	Recov-ered.	Result stated.	Recur-rence.	No recurrence up to				
				1 yr.	2 yrs.	3 yrs.	4 yrs.	5 yrs.
Sarcoma and its combinations, including chondrosarcoma . .	26	18	10	1	1	1	1	2
Chondroma, simple . .	7	4	2	1	...	...	...	1
Myxochondroma . .	1	0						
Carcinoma . . .	2	1	...	1				

A study of this table shows very little difference in mortality between the extrapleural operations and those which involved the opening of the pleural cavity. This seems on

---

* This case has been by error included; it should be in the table of extrapleural operations.

† In this case (Gibson's) recurrence took place only after six years, causing death.

first reflection a very remarkable circumstance, and all the
more so since in cases of the second series the diaphragm was
involved in the disease and included in the operative intervention, and in some cases portions of the lung were actually
excised. A few observations here will easily set the matter
right :

Of the non-penetrating operations, 12 or 46 per cent., were
done prior to 1873, hence were pre-Listerian. They were done,
too, at a time when surgical technique, apart from its present
antiseptic or aseptic accompaniments, was yet in comparative
infancy.

In the second series or operations involving the opening
of the pleural cavity, on the other hand, of the 52 cases
reported upon 9 only were done before 1878, or 17 per cent.,
and of these the deaths of two cases were probably (confess-
edly in one case, that of Langenbeck) due actually to the
antiseptic used in the early days of the Listerian struggle
with wound infection—a period in surgery when only the
good effects of carbolic acid were believed in, its intoxicant
power not being recognized, even though black urine continu-
ally warned of the patient's jeopardy.

This case has been rightly styled by Leonhard Meyer
(*Inaug. Diss.*, Erlangen, 1889) as an " Opfer der Antiseptik."

This argument is further emphasized when we study the
recurrences in the two classes of cases. Inspection of the
tables of recurrences does not at once show marked difference
in the two series, but this is to be partly accounted for by
the lack of data in a large number of cases. One thing, however, is apparent, that in the intrapleural operations more
cases are credited with long survival after operation than in
the other series. This can only be accounted for by the fact
that in the cases prior to 1878 imperfect technique was more
conspicuous, and operations were not undertaken with that
boldness which is, pre-eminently in this field of surgery, the
assurance of thoroughness in operative work. It was not so
much a difference in operative experience in the men doing
the work, but rather a timidity born of the consciousness

that the operative field was new, comparatively untrodden, and the surgical world had not been convinced by the operative success of Richerand in 1818. Witness the fact, drawn from a study of these tables, that the second intrapleural operation (that of Sedillot) was not done until 1855, fully thirty-seven years after Richerand's famous case, unique, therefore, for the first half of the century.

A study of the tables reveals no striking difference as to mortality in either series between the malignant and the so-called innocent class of tumors—enchondromata.

Further, there seems almost as little difference in the tendency to recurrence—an observation which will naturally surprise those who have been in the habit of regarding the chondroma as benign. These tables show anything rather than benignancy. I desire to insist with some emphasis on this point. The extremely slow, therefore insidious, growth of these chondromata, lasting in some cases for years (witness the cases of Kappeler, twenty-five years; Morell-Lavellée, thirty-five years), and developing without pain, encourages individuals so afflicted to delay seeking the surgeon until the tumor by its size and weight becomes a burden and interferes with occupation ; then, often, it is too late for radical extirpation, for extension of growth, directly or by metastasis, may have taken place, as in Roswell Park's case (Case XXVIII., second series).[1]

A strong argument in favor of early operation is that under circumstances not at all understood (perhaps an injury may have had some influence) these benign tumors take on malignancy due to their transformation in parts into sarcoma,

---

[1] A good example of metastasis of chondroma is that reported by Richet (Gaz. des Hôp., 1855, p. 373) of a man thirty-four years of age, operated on for a tumor (pronounced by Broca a chondroma) of the scapula, size of a child's head, noticed four years , stationary three years ; autopsy two weeks after operation discovered on posterior surface of lung of same side a tumor size of a hazel-nut, and about thirty tumors from size of a millet seed to that of a nut scattered in both lungs. It is possible that at this early day the evidences of transition into sarcoma may have been overlooked. A good example of extension by direct growth is the following : Paget, in his "Surgical Pathology " (1853), reports an instructive case of chondroma of the heads of the ribs, the tumor projecting itself into the vertebral canal, producing complete paraplegia and death.

their rate of growth, and tendency to extension and metastasis being much accentuated.

Unquestionably our statistics would be much improved, both as to immediate and remote results, if the operative intervention could be undertaken at a time when radical operation would not involve the opening of the pleural cavity.

The enchondromata continue rarely as pure forms, showing, after a time in the greater number of cases, combinations which change their character and hasten their growth.

A study of the cases here presented will abundantly establish the malignant tendencies of chondromata. These qualities of malignancy may be thus enumerated :

1. Tendency to sarcomatous change.
2. Extension by contiguity (infiltration).
3. Metastasis to inner organs.

Chondroma, moreover, seems to be one of the more frequent tumors in this region, although this may not clearly appear from the tables, only thirty-two of the seventy-eight cases, or 41 per cent., having apparently begun as chondroma.

As to this comparative regional frequency, the carefully prepared statistics of Weber in his thesis on the subject (see Schläpfer von Speicher) may be quoted. (Weber collected 237 cases ; to these Schläpfer added 23.) Of 260 cases in bones generally, only 8 were found in the ribs—that is to say, about 3 per cent. of all chondromas are located in the ribs.

It may be true, as Senn says, that it is not improbable, when a chondroma has invaded adjacent tissues, or metastasis has occurred, that there has already been transition into sarcoma. "The exciting causes," says Senn (*Tumors*, p. 84), "in effecting a transition of a benign into a malignant tumor are such local and general influences as transform mature cells into embryonic cells, and which at the same time render the surrounding tissues more passive to cell infiltration."

The local causes establishing this post-natal embryonic matrix, which is the starting-point of malignancy, may be thus enumerated (after Senn) : Injury, prolonged or repeated irritation (rheumatism, in one case), and incomplete removal of the benign tumor by excision or by cauterization. Age, of course, will have some influence in this transformation. Enough has been said to urge the early removal of benign tumors when operation is easy and safe and before such transformation has taken place.

There is a time in the life of all tumors when they are local and can be radically removed. Leonhard Meyer laments the late period at which operation is usually sought, and expresses the hope, naively, it is true, but with reason, that the laity will understand the importance of reporting early, so that the best results may be obtained.

I have spoken of the influence of irritation in the transformation of tumors. It might be interesting to note here the effect of traumatism. Senn has some instructive remarks on this subject. While traumatism alone is incompetent, acting primarily, to produce a tumor, still it " in exceptional cases may and does act as an exciting cause in the growth of a tumor by diminishing the physiological resistance of the injured tissues, or by causing irritation or inflammation in the immediate vicinity of a pre-existing tumor-matrix; or in more exceptional cases it furnishes both essential conditions for tumor growth—a post-natal matrix of embryonic cells and a diminution of physiological resistance in the immediate vicinity of the new matrix." (Senn on *Tumors*, p. 69.)

In my cases trauma is stated in only five cases of Series I. and in only fourteen of Series II. Of course, the evidence in many cases may be deemed inconclusive, but the fact is well attested and is accepted by most pathologists.

This statement applies not only to sarcoma, but to chondroma also. The truth of the statement as to chondroma is well illustrated by the experiment of Heyfelder, quoted by Schläpfer von Speicher from Virchow's *Archiv*, 1858, Band xiii. A four-year-old dog got a blow on the left side. After death a chondroma seven by four and one-half inches on the outer aspect, and another three and one-half by three inches on the inner surface were found. Instances in man of chondroma in various situations developed by trauma could easily be multiplied.

As to carcinoma of the ribs, it is doubtful if it ever occurs in the primary form, being usually consecutive to cancer of the breast.

The influence of age and sex are displayed in the following table :

AGE.

		First Class.	Second Class.
1.	Decade from 1 to 10 years of age .	. . 1 (Case 6.)	0
2.	Decade from 10 to 20 years of age .	. . 3 (Cases 15, 19, 26.)	7 (Cases 14, 16, 29, 40, 44, 45, 48.)
3.	Decade from 20 to 30 years of age .	. . 4 (Cases 10, 18, 22, 27)	11 (Cases 9, 15, 24, 26, 28, 32, 39, 49, 50, 51, 22.)
4.	Decade from 30 to 40 years of age .	. . 7 (Cases 4, 7, 9, 12, 16, 17, 24.)	12 (Cases 6, 12, 13, 23, 25, 27, 30, 31, 35, 42, 7, 33.)
5.	Decade from 40 to 50 years of age .	. . 3 (Cases 3, 20, 23.)	8 (Cases 1, 11, 17, 21, 10, 36, 46, 48.)
6.	Decade from 50 to 60 years of age .	. . 3 (Cases 13, 21, 25.)	2 (Cases 34, 38.)

SEX.

	Series I.	Series II.		
Male	. 18	34	=	52
Female.	. 8	14	=	22
Not stated	. 0	4	=	4
Total cases .	. .	.		. 78

DIAGNOSIS. The diagnosis of these parietal tumors is an interesting subject, but can only be briefly referred to here. Mistakes may be easily made, but if the history of development be given due weight, error is not apt to occur. As showing the difficulty of arriving at a conclusion without a consideration of the history of development, I mention the cases of Heyfelder (reported by Hermann Schüster, *Inaug. Diss.*, Erlangen, 1851), of a man, aged thirty-three years, who in January noticed a painless, movable, bean-sized tumor, which grew in five weeks to the size of a hen's egg, and by the end of March was as large as a female breast. It was found at autopsy to be a tumor filled with blood, due to erosion of an intercostal artery connected with caries of the fourth rib.

For the operative diagnosis of the tumor it is only necessary to determine that the tumor can be removed; in other words, that it is not a thoracic aneurism, nor other tumor of mediastinal origin, making its way through the chest wall.

A case illustrating the difficulty of diagnosis in some rare instances is found reported in the London *Lancet*, 1859, vol. i. p. 489. The specimen was sent to the Pathological Society of London by Mr. Sharpley. It was removed from the body of an old man who had suffered with it for twenty-five years, and had been seen by some of the most eminent surgeons of London and the provinces. Some considered it a fibrous growth; others, like Mr. Liston and Mr. Key, thought it an aneurism. It was very large, and grew both within and without the chest, the two portions being connected through a hole in the sternum. It had large blood-vessels, and the man's diary stated that there had been a bruit.

The beautiful skiagraphic plates to be found in Murphy's article on the " Surgery of the Lung," in the *Journal of the American Medical Association,* July 23 to August 13, 1898, show that the X-rays are destined to render much valuable service in the surgery of the chest. The position of the heart, the diaphragm as it moves in respiration, the gradual expansion of the lung as the retained pleural air absorbs, can all be studied to a certain extent with the fluoroscope, and still better in the skiagrams, and as the apparatus becomes improved the delicacy of its definition will become more and more apparent. The prize essay on "Skiascopy of the Respiratory Organs," just published in the March number of the *Philadelphia Monthly Medical Journal,* indicates already some material advance in the near future.

The percentage of hemoglobin should be determined, a decided fall in amount indicating a tendency toward malignancy or the imminence of recurrence. See case of Mikulicz (XXXV.).

TREATMENT.   At the outset the question should be considered whether these tumors, having reached a certain stage
in development, had better be removed or left alone.   As to
carcinoma, the question is not so difficult, since these tumors,
being secondary in character, do not offer the same justifica-
tion for a formidable operation.   While Richerand's famous
case was the first intrapleural operation, and was in this sense
an epoch-making event in the surgical world, and would have
been accepted without question by surgeons as a justifiable
operation if the character of disease had made recovery possible, still the rapid recurrence of the disease and the death of
the patient only one month after having passed through such
a trying surgical ordeal without an anaesthetic, caused the hand
of the boldest surgeon to pause, while he reflected whether it
would not have been better to have let the poor man die by
the hand of God than to have subjected him to the torments
of three ineffective mutilating operations.

Without doubt surgery was checked in its progress, for not
another case, even operable in character, was done until Sedillot successfully attempted it in 1855; but, incontestably,
Richerand deserves the credit for the demonstration that the
opening of the pleural cavity can be successfully done.   The
seed had remained, and after many years, fortunately for this
branch of surgery, blossomed into fruit.   Few surgeons now
would have the boldness to repeat under the same circumstances the experiment of Richerand,[1] notwithstanding the
brilliant success of Schede, whose patient was alive and well
one year after an extensive operation for recurrence after
excision of a breast carcinoma; and of Finck, whose patient
lived three years and eight months after an extirpation of a
mammary cancer, in which he wounded the pleura, giving
rise to a severe pyopneumothorax.   In any case of secondary
tumor one must, at all events, be sure before operating that
there is not elsewhere another tumor.   Regarding the difficulty of ascertaining whether a cancerous tumor is primary

---

[1] According to Plitt (Inaug.-Diss., Berlin, 1890), Richerand afterward declared that
he would not again do this operation for cancer.

or secondary, Roulliés (*Thèse de Paris*, 1887–1888) quotes a communication made to him by Nicaise: "I have seen," says that able surgeon, "several cancers of the sternum, all of secondary origin, but once I did not find at first the primary cancer; and I asked myself if I were not in the presence of a primitive tumor, and if it were not proper to do an operation. In the meantime the patient fell unaccountably and sustained a spontaneous fracture of the neck of the femur, due to a cancerous nodule, confirmed by autopsy. The primary disease was a renal cancer, absolutely inactive, without apparent alteration of the urine."

If, however, a tumor be demonstrably primary in character, and there be reasons for believing that it has invaded no structure which cannot be removed, then an operation is indicated, and may be done if the condition of the patient shall justify it. A few carcinomas secondary to mammary cancer, then, and fibromas, chondromas,[1] sarcomas, and, exceptionally, gummata of the ribs and sternum, will fall into this category. An interesting problem in surgery was broached by Küster (see sternal cases), which has not yet been definitely settled, whether, as in his case, the surgeon would be justified in doing a costal or sternal resection for gumma which had resisted a thorough six weeks' course of iodide. In his case, although there was absolutely no history of syphilis and no other signs of gumma, a syphilitic tumor was suspected and the iodide tried thoroughly without effect on the neoplasm, whereupon extirpation was done, and its gummatous nature established on examination. The therapeutic test had led him into error, it is true, but its failure after adequate trial had, I believe, justified the operation. I learned through a personal communica-

---

[1] When one looks superficially at the question he is tempted to let these tumors of the chondroma type severely alone. Thus, of the fifteen cases tabulated by Schläpfer von Speicher, three had lasted 1 year and under (two were operated on with cure), two had lasted three years, two four years, two five years, two six years, one eight years, one twenty-five years, one thirty-five years, and one several years, but it must not be forgotten that chondroma of the ribs is very prone to undergo metamorphosis and to form various combinations which hasten their growth, and these changes are very prone to occur under the influence of trauma. (See Section on Transformation of Benign into Malignant Tumors.)

tion of one of the Fellows of the Southern Surgical and Gynecological Association of an equally interesting case of opposite character, in which, in the belief that a sternal tumor was a gumma, the antisyphilitic medication was persisted in for a considerable period, but without effect. The autopsy showed osteosarcoma of the sternum, which would have been amenable to operative treatment.

The tumor having been decided to be operable in character, the surgeon must determine beforehand how far he will carry his operative attack in his endeavor to get beyond the limits of the disease. Shall he stop at the pleura, or, having gone into the cavity, let the diaphragm be his *noli me tangere?* or will he proceed further, stopping at nothing short of the heart or the whole of one lung? These are questions difficult to answer, but I shall attempt to indicate the chief, at least, of the factors which aid in the solution of the problem.

It might be wisest, first, to consider the causes of death in the cases here reported. It will be profitable to consider only the second series.

These may be ranged under the following heads :

Shock : Weinlechner (14), Hahn (26), Park (28), Helferich (49)	4
Shock and hemorrhage : Koenig (39)	1
Shock and pneumothorax : Sheild (41), Quénu (44)	2
Hemorrhage and pneumothorax : Tietze (12) . .	1
Hemorrhage and Sepsis : Heyfelder (4) . . .	1
Chronic sepsis : Israel (8)	1
Antiseptic irritation (capillary bronchitis) : Leisrinck (13) .	1
Carbolic intoxication : Langenbeck (7)	1
Pyopneumothorax : Schuh, (3) Wattmann (10) . .	2
Doubtful	2
Total	16

The ages of these cases ranged from ten years (Sheild's case) to sixty (Schuh's).

The cases showing shock were in order, beginning at the top—thirty-seven, thirty-three, ten, fifty-five, ten, and forty-nine years of age. Of the remainder the oldest case was that of Schuh, sixty years, dying in two days of pyopneumothorax. The others ranged between thirty-two and forty-six years of age.

Sex seems to have had even less influence in determining the fatal result. Thus, of the sixteen fatal cases only two were females (Langenbeck's and Israel's). The condition of the patient at the time of the operation is specifically stated in very few cases. Thus, Case XIII. (Leisrinck's) is said to have been pale and dyspeptic; Case XXVI. (Hahn's case), exhausted by repeated operations; Case XXVIII. (Park's case), debilitated by repeated operations for recurrent growths.

Even in cases which did not die the condition was at times so threatening that it will be very profitable to study all the cases in groups, in order to arrive, if possible, at some satisfactory solution of the phenomena observed.

PNEUMOTHORAX, HEMORRHAGE, AND SHOCK. Without doubt the first is the most formidable complication attending operations which open the pleural cavity. Hemorrhage, whether from the intercostal arteries or the internal mammaries, may be prevented or controlled by multiple cartilage or rib resections and ligation, as done long ago by Langenbeck and Richter, or by clamp-forceps, suture-ligature, or tamponade, if the ordinary method of ligature cannot be carried out. It is worthy of remark just here that the bleeding from the intercostals in the course of the excision of a parietal tumor is often surprisingly little.[1] In my second case, for example, not a single intercostal vessel had to be tied. The lung itself bleeds little; suture-ligature and ligature of even the large vessels accompanying the bronchi, or putting on two large clamps, as was done by Helferich in his last case, may be justifiable in certain grave emergencies.

But the greatest danger, and the one most difficult to deal with, is pneumothorax. We shall now proceed to study this.

STUDY OF THE PHENOMENA OBSERVED ON OPENING THE PLEURA. *Classification of the Cases according to the Severity of the Manifestations.* The following cases are thrown

---

[1] The absence of hemorrhage was one of the points made against Richerand by Nicod. He assumed because no intercostal bled, that this was the result of the progressing disease, enveloping and obliterating them; but this may not necessarily have been the case at all.

ont because of insufficient data: Cases II., III., IV., V., VI., VII., VIII., X., XIX., XXVI., XLV.—eleven in all—except in so far as they may illustrate points mentioned in the histories.

The remaining forty-one cases may be grouped as follows:

1. Cases marked by little or no disturbance when the pleura was opened, either specifically so stated or whose clinical histories justify the assumption that the disturbances were slight: Cases of Tietze (XII.), Desguin (XXII.), Tietze (XXIX.), Alsberg (XXXIV.), König (XXXVI.), Zarubin (XXXVIII.), Sheild (XLI.), Marsh (XXXIII.), Alsberg (XXVII.), Trendelenburg (XX.), König (XL.), Tietze (XXI.), Heineke (XXIV.), Bardeleben (XXXI.), Mikulicz (XXXV.), Helferich (XLIX.).

2. Cases marked by only moderate disturbances when the pleura was opened : Cases of Cassidy (IX.), Maas (XVIII.), König (XXIII.), Park (XXVIII.), Bardeleben (XXXI.), Vautrin's second case (XXXVII.), Thompson (XLVII.), Faure (XLII.), Gauthier (LI.), Parham's second case (LII.).

3. Showing quite stormy manifestations, even threatening life : Richerand (I.), Fischer-Kolaczek (XI.), Leisrinck (XIII.), Weinlechner (XIV.), Krönlein (XV., second operation), Humbert (XVI.), Müller (XXV.), Vautrin's first case (XXX.), Witzel (XXXII.), König (XXXIX.), Karewski (XLIII.), Quénu (XLIV.), Bayer (XLVI., both), Doyen (XLVIII.), Parham's first case (L.).

THE NATURE OF THESE DISTURBANCES. In this discussion I have drawn freely upon the valuable work of Gerulanos. In all the cases the respiration was interfered with to a certain extent when the pleura was opened, being somewhat accelerated.

In the cases of Leisrinck (XIII.), Weinlechner (XIV.), and Müller (XXV.) the breathing and pulse were much inhibited. The patients were severely collapsed in seven cases, the pulse being small, scarcely perceptible in them all. Though in the fatal cases this is not so apparent, still in a

general way it may be admitted that the age, the condition
of the patient, and the existence of complications, such as
affections of the heart (König, XII., XXXIX.) and tuberculosis of the lung (Pfeiffer), will influence the termination.

The kind of tumor, its size, its extent, and its relations to
the lung and pericardium will also be seen to have some
influence.

A reference to the table will convince one that the greatest
factors are the size of the opening and the time it remains
open. Thus it will be seen that in Cases XI., XIII., XIV.,
and XV. the pleural defect was as large as a man's or a child's
head. In all the other cases marked by stormy symptoms the
opening was as large as the hand or as the palm of the hand,
and in all these cases the tumor was large and its removal
required some time, the patients all, with the exception of
Cases XI. and XXXIX., being under forty-five years of
age; all except XI. and XV. being male, and all being in
excellent condition.

In the case of smaller pleural defects or merely tears the
manifestations were not severe, since the pneumothorax was
easily controlled by gauze tamponade. In some few cases
(XXIV. and XXXI.) adhesions prevented the development
of collapse. A specially remarkable case is that of Mikulicz
(XXXV.), where, in a fifty-four year old patient, a large
tumor was removed with excision of a piece of the diaphragm
with almost no disturbance; and that of Maas (XVIII.)
showing only a slight slowing of the pulse and respiration.
These were both on the left side, low down and behind. In
cases of the first class adhesions explained the absence of
severe symptoms (König, XXI.), or the tumor was small.

OPERATIONS ON THE LUNG AND DIAPHRAGM. Of the
cases of the third class operations were done on the diaphragm
in XIII. and XVI , and on the lungs in XIV., XV., XXV.,
and XXVIII. The case of Helferich (XLIX.), in which
actually two lobes of one lung were excised in a boy, aged
fifteen years, with slight pneumothorax, would seem to indicate the correctness of the assertion of Gerulanos that in the

cases now mentioned the disturbances were due less to the
fact that the lung or diaphragm was involved in the operation
than that the operative work took more time and kept the
pleural wound open so much the longer. This is further
shown by Case XXXI. and König's sternal case. It is
well, if prolonged intrathoracic work must be done, to keep
the cavity as far as possible packed with moist gauze.

THE SIDE OF THE THORAX INVOLVED. With rabbits
(Gerulanos), right-sided pneumothorax is immediately fatal,
because the left lung is a third smaller than the right. The
oxygen-content of the aortic blood amounts, when one pleural
cavity is opened—if the left pleura to 76 per cent. ; if the
right pleura to 58 per cent.

Of the cases in the third clinical class (showing stormy
manifestations) the lung involved was : the right in nine
cases; the left in five cases.

Of four cases with great defect, yet little disturbance, in
three (XVIII., XXII., and XXXV.) the left, and in only
one (XXIX.) the right was involved; and it is probable that
adhesions partly explain this, for the age of the man was
rather in favor of bad symptoms.

The explanation in this right-sided and left-sided pneumothorax seems to be twofold :

1. The lung thrown out (greater disturbance when the right).

2. As to which way the mediastinum is pushed by the
pneumothorax.

Right-sided pneumothorax pushes the mediastinum toward
the left, compressing the venæ cavæ and the thin-walled right
auricle, while left-sided pneumothorax exerts its pressure on
the more resisting right *ventricle*.

In the first case the sucking up of the venous blood from
the cavæ into the right auricle is seriously interfered with.

In addition to these anatomical considerations must, of
course, be taken into account the hemorrhage, the narcosis,
and the circumstances surrounding the patient. Of great
significance is the elimination of one lung from the work of
respiration.

Reflex irritation, however, seems to be of more importance than the simple compression of the lung. Quénu and Longuet lay great stress on this point. The effect on the heart is partly reflex, partly due to direct pressure on the heart and great vessels (especially the great veins and right auricle) in right-sided pneumothorax. Witzel attaches some importance to the flexion of the great vessels. According to the investigations of Sackur on rabbits, the circulation through the collapsed lung is not diminished, rather, indeed, is increased.

This reflex effect is likely to be prolonged or intensified by pleural irrigation. Where the wound cannot be closed afterward, as in Weinlechner's and Helferich's cases, the fatal result may be largely attributable to this reflex disturbance.

Quénu and Longuet, in their very valuable article, give the following explanations of the varied manifestations of pneumothorax in different cases. I have grouped them, for greater convenience of reference :

1. Pulmonary retraction may be incomplete, on account of adhesions. These may be complete enough to shut off the pleural cavity, or only partial, the location of the adhesions being of great importance; or the lung may not contract, because less elastic from œdema, hepatization, or sclerosis, etc.

2. The effect of reflex irritation is of the greatest influence, as was shown in the case of a colleague of Quénu during his interneship, who, instead of making a vacuum in his syringe for aspiration, injected air, with the instant death of his patient. *The wide opening of the pleura by letting a large amount of air suddenly into the chest cavity intensifies the reflex effect.*

A difference must be made between those cases which breathe calmly throughout the narcosis and those which show stertorous breathing from the start. In the former case when the pleura is opened the in-rush of air is much more effective in giving rise to the reflex irritation.

THE EFFECT OF DRAINAGE. The question of drainage is one requiring earnest consideration. One must weigh the necessity of providing for the discharge of the secretion from the irritated pleura with the necessity of preventing secondary pneumothorax. Quénu and Longuet suggest a middle course by the use of an apparatus similar to Bigelow's suction apparatus for the bladder after litholapaxy, so as to get drainage without admission of air. I believe if drainage is adopted other than into a vacuum, gauze should be preferred to tubes, as we know how a little piece of gauze will control pneumothorax and will at the same time permit of sufficient discharge. A study of Maas's case shows how, with satis-factory asepsis, the wound may be safely closed without any provision for drainage beyond leaving a small, dependent corner of the wound unsutured. Animal experiments accord with clinical observations, that the air is gradually absorbed. A study of the cases like that of Maas shows that the air in the cavity is rapidly absorbed, in his case by the fifth day. Even with two drainage-tubes (Krönlein's case) the pneumothorax disappeared rapidly. See also cases of Leisrinck, Müller, Alsberg (XXIV.), and Bardeleben (XXV.). Bouveret and Schede, says Gerulanos, have called attention to the valvular action of a tightly applied dressing, and Aufrecht broached the theory that the collapsed lung will expand in every inspiration when the chest opening is smaller than a main bronchus.

THE LIMITS OF OPERATION. Although the result was good in Humbert's case, he was so impressed with the dangers and difficulties of the operation when the diaphragm was involved in the disease that he counselled deciding against operation if it were possible to determine beforehand that the diaphragm would in all probability have to be attacked with the knife.

Let us examine the cases and classify them so as to see what our limitations are :

Costal operations, extrapleural (Case 20), in which the pericardium
    was opened ; there was immediate fall in the pulse, but he
    recovered 1
penetrating, pericardium opened 0
diaphragmatic adhesions separated (Cases 12 23 . . 2
adhesions between lung and tumor separated Case 25, Park ,
    died 1

A consideration of these results and the phenomena observed in the course of operation justifies the conclusion that the trouble rather arises from the prolongation of the operation by reason of the intrapleural manipulation than on account of the severity of the procedures.

The remarkably successful experiments of Biondi on animals, in 1882, in which he extirpated the right lung in twenty-three, with twelve recoveries; both apices in three, with three recoveries; middle lobe in one, with one recovery, and the almost equally good results of Zakharwitch (Murphy's paper), demonstrated that, at least in animals, there was almost no limit to the possibilities in operations on the lung. The recovery of Milton Antony's patient (in Georgia) in 1821, in whom he had resected two ribs and scooped a tumor out of the lung (this was the operation which Dieffenbach criticised afterward so severely), and the equally successful case of Péan, in 1861 (Murphy is wrong in stating that this was a tumor of the ribs; Paget quotes the case from the original article as one simply of tumor of the lung), were the first pneumectomies on the human. Since these the lung has been attacked in a considerable number of instances with success. Witness the case of Tuffier in 1891, done with the stripping of the parietal pleura; Mr. Lowson, of Hull, in which he resected the apex of the right lung after preliminary injection into the pleural cavity of air through a trocar, connected with a Junker's bottle and bellows,[1] and the cases referred to in this paper. These cases have brought the lung into the field of operative surgery, and though no case is on record where the whole lung, nor, indeed, a whole lobe, has been removed in man with success, yet the results in animals certainly seem to indi-

---

[1] Lowson, not Lawson, as some have it. Brit. Med. Journal, June 3, 1893, p. 1152.

cate the possibility in the future of more extensive work. The excision by means of the V-shaped incision of Schmid (1881), with sewing over of the lung stump, so as to cover in the bronchus, in conjunction with the Fell-O'Dwyer apparatus for forced respiration, and the proper closure of the parietal wound, will yield better results than was obtained by Helferich in the case reported in this paper (Case XLIX.), in which he resected two whole lobes of one lung with only slight circulatory and respiratory disturbances, and without any artificial respiration. He utilized a portion of the lung in covering the root-stump. Murphy has demonstrated by experiments on dogs that the root of the lung may be grasped not only without bad effect, but actually with decided improvement of all the symptoms due to the pneumothorax. In a dog, too, which lived eighteen weeks after extirpation of the three lobes of the right lung, he showed that there was definite closure of the large bronchus tied before the lung was cut away.

Finally, Murphy has given us very satisfactory physiological argument to justify our expectation that the patient may be able, at least during the operation, to dispense with one whole lung.

"Our extreme respiratory capacity," he says, "is 3358 c.c. In an ordinary exchange it is 114 c.c. at rest to 360 c.c. in exercise; therefore, the extreme capacity is equal to at least ten times the ordinary exchange. Pathologically, we recognize that with one lung entirely disabled by pleuritic effusion, and the capacity of the other lung greatly diminished by displacement of the mediastinal septum, the respiratory change is still able to sustain life. These physiological and pathological facts prove that with our surgical technique sufficiently advanced an entire lung may be removed, as the body can dispense with it." "The accepted theory," he continues, "is that the exchange between lungs and blood is effected by physical and chemical means, the most important of which is diffusion. The surface through which this diffusion takes place is ninety square metres, and through this there are diffused 300 c.c. of carbon dioxide and about the same quantity of oxygen per minute in ordinary respiration."

"The question which concerns the surgeon in this relation is how much of the ninety square metres of this surface can be dispensed with suddenly, and how much can be dispensed with slowly, the former by amputation or compression of the lung, the latter by disease processes. Clinical experience shows that one lung may be suddenly rendered entirely functionless without materially embarrassing respiration, as in pneumothorax and hematothorax, as I have demonstrated experimentally and therapeutically by nitrogen gas injections into the pleura." A case particularly illustrating this point is that of Heineke (Case XXIV.), in which, although the right lung was practically eliminated from the work of respiration, there was no noteworthy disturbance of either circulation or respiration.

We may then feel justified when we start out to remove a tumor of the thoracic wall, in doing a radical extirpation of the disease, provided that it does not carry us beyond the capacity of the individual patient to endure the operative attack.

SUGGESTIONS FOR THE PREVENTION OF PNEUMOTHORAX.

1. Early operation, before the tumor is of great size.

2. Preliminary adhesions (see Billroth's and Gauthier's cases).

3. Substitution of temporary hydrothorax for pneumothorax. (See Witzel's case, XXXII.)

4. Injection of sterilized air (Lowson, *British Medical Journal,* June 3, 1893).

5. Tuffier and Hallion's intubation method.

6. Quénu and Longuet's intrabronchial tension, with the aid of compressed air to neutralize the mediastinal vacuum.

7. Doyen's method.

8. Fell-O'Dwyer apparatus (see Parham's case, LII.).

9. Stitching up the lung to the margin of the thoracic wound, as practised by Tuffier, Bayer, and myself. Quénu objects to this as causing emphysema, but I believe the objection of little moment.

10. Obturation of the wound of the pleura, (a) by the finger; (b) by a compress; (c) by suturing the flap. Some-

times by taking a flap from a distance with a pedicle (as Vautrin did in his first case) the wound may be closed.

11. Freeing the parietal pleura by dissection, sinking it into the chest, and uniting it by sutures with the visceral layer, as practised by Vautrin in his second case (XXXVII.).

12. The patient may be directed at the termination of the operation—when the wound has been closed, except where a small tube is inserted—to take a deep inspiration to force the air out of the pleural cavity; at this moment the tube is quickly withdrawn and the wound hermetically closed.

This was done in Marsh's case (XXXIII.).

13. The air may be gotten rid of, partly at least, by aspiration, as in my second case.

THE TREATMENT OF SHOCK AND HEMORRHAGE. In this class of cases the treatment of shock will not differ materially from its treatment in other operations. Saline infusion will render conspicuous service in chest surgery. To get the best results it must be given intravenously, and must always be given as hot as can be borne, say 115° to 120°, especially when given for shock.

It strikes me that the new surgical table of Pyle, described in a recent number of the *New York Medical Journal*, equipped as it is with every provision for an emergency, and especially its arrangements for keeping the patient warm, will prove particularly useful in intrathoracic surgery.

I cannot leave this subject without again paying tribute to the Fell-O'Dwyer apparatus for forced respiration.

Matas has, in his paper referred to, given a history of the evolution of this admirable apparatus, and has indicated its wide range of usefulness in chest surgery. The assistance rendered me in my second case was so striking that I can without hesitation indorse every word that has been said in its favor. Indeed, so imbued am I with its great value that I believe no surgeon now would be justified in attempting a thoracic resection without having the Fell-O'Dwyer apparatus at hand. I believe it will revolutionize this field of surgery, making possible operations in the chest that would otherwise be clearly too hazardous to be justified.

RESECTION OF THE STERNUM FOR TUMORS.

No.	Reference.	Sex and age	Nature of tumor.	Operation.	Complications.	Result.	Remarks.
1	Holden, Brit. Med. Journ., 1878, ii. p. 358.	F. 52	Sarcoma of gladiolus partly calcified.	Imperfect removal of tumor only to posterior surface of the bone.	....	....	The more successful methods of modern surgery were hardly then in vogue.
2	König, Centralbl. f. Chir., 1882, No. 42.	F. 36	Sternal sarcoma.	Part of manubrium, the gladiolus and the ribs (six) from second downward (left), and ends of cartilages on the right.	Both pleuræ and pericardium opened; both internal mammaries tied.	Recovery ; Died 2 yrs. later from recurrence in lungs.	Only slight respiratory trouble ; great frequency of pulse ; duration of operation 3 hours ; collapse slight at termination.
3	Küster, Berlin klin. Woch., 1882, p. 127.	M. 30	Gumma ot gladiolus.	Right half of gladiolus and 3d and 4th ribs.	Right pleura opened, size of 10-pfennig piece.	Recovery.	
4	Pfeiffer, Beitrag. z. Kennt d. Sternal Tumoren. Halle, 1884.	M. 45	Sarcoma of gladiolus.	Gladiolus and 2d, 3d, and 4th ribs ; right and left side.	Right pleura opened and pericardium exposed.	Died 6th dy. pericarditis, hemothorax and pleurisy ; fresh tubercles in lungs.	Other tumors found at autopsy on 4th rib, pubes, in aortic glands, liver, and right kidney.
5	Bardenheuer, Deutsch. med. Woch., 1885, xi. p. 688.	F.	Fibroid of manubrium.	Manubrium, two-thirds of clavicle, 2d and 3d ribs.	....	Recovery.	Tumor extended to larynx, gladiolus, and both clavicles.
6	Ibid.	M.	Sarcoma of clavicle.	Manubrium, two-thirds of clavicle, 1st and 2d ribs.	Int. jugular torn ; tied rt. innominate, subclavian, int. and ext. jugular veins.	Recovery.	
7	Ibid.	...	....	Manubrium.	....	Recovery.	No other details given.
8	Ibid.	...	....	Manubrium.	....	Death.	No other details given.
9	Ibid.	...	....	Manubrium.	....	Death.	No other details given.
10	Ibid.	F.	Retrosternal fibroid.	Manubrium, part of gladiolus, 1st and 2d ribs, and inner end of clavicle.	....	Recovery.	
11	Ibid.	...	Retrosternal fibroid.	" " "	....	Recovery.	
12	Jaenel, Resectio sterni. Inaug. Diss., Erlangen, 1887.	F. 12	Sarcoma of manubrium.	Manubrium, 1st and 2d ribs.	Retro-sternal tissues involved, and cauterized with nitric acid and cautery.	Recovery.	No return after a year.
13	Dudon, Jour. de Med. de Bordeaux, June 1, 1890.	F. 28	Enchondroma of gladiolus.	Manubrium, part of gladiolus, and 1st and 2d ribs.	Internal mammary exposed but not injured ; suppuration of wound.	Recovery.	Tumor first remov'd by oper. 27 mos. before : 3 small nodules removed 8 mos. later from pectoral and sternocleidomastoid ; ascribed to bringing elbows forcibly together behind her back.

No.	Reference.	Sex and age	Nature of tumor.	Operation.	Complications.	Result.	Remarks.
14	Mazzoni, in Dudon's paper.	M. 55	Myxo-sarcoma of gladiolus	Gladiolus and 2d, 3d, and 4th ribs.	None.	Death 15th day from pneumonia	
15	Graves, Med. News, Mar. 4, 1893, p. 241.	F. 44	Sarcoma of gladiolus.	3 x 3½ inches gladiolus and 2d to 5th ribs.	None.	Recovery.	Right breast removed for carcinoma 22 months before.
16	Doyen, Arch. Prov. de Chir., 1895, iv. p. 633.	M. 37	Sarcoma of manubrium.	Manubrium, parts of both clavicles, and 1st rib.	None.	Recovery.	Rt. and lt. brachio-cephalic vein disclosed; lungs and heart seen; but neither pleura nor pericardium opened.
17	Mynter, Annals of Surgery, 1891, xiii. p. 96.	F. 20	Melano-sarcoma of gladiolus.	Gladiolus (6 sq. in.) from 3d nearly to 6th ribs, and 3d, 4th and 5th ribs.	None.	Recovery.	Weight bet. ¹₂ and ³₄; removed, also glands from both axillæ and sub-clavian glands; pericardium exposed. Ascribed to injury from handle of broom. Dr. Mynter kindly writes me that patient died in Ireland a year later, probably from recurrence.
18	Keen, Med. & Surg Reporter, March, 1897, vol. lxxvi.	F. 28	Sarcoma manu-brium and ster-nocleidomastoid.	Manubrium, ½ clavicle, 1st rib, and ½ ster-nocleidomas-toid.	Suppuration of wound; drainage by Cathcart's method.	Recovery.	No recurrence after 15 months.
19	Ibid.	F. 44	Carcino-ma of manu-brium and gladi-olus.	Part of the manubrium and the gladi-olus and 2d and 3d ribs.	Carcinoma of both breasts.	Recovery.	Lt. breast removed March 1, 1891; re-current nodule in scar May 20, 1893; right breast and parts of sternum Mar. 22, 1894; re-currence and death Aug. 1894.
20	Weinlechner, Bericht d.k.k. Krnkanstl. Rud. Stift. in Wien (1871), 1873, p. 124.	M. 58	2 tumors; primary, chondro-ma; sec'dary, colloid sarcoma; hard tumor 50 years' duration, soft tum. 3 years.	Soft tumor re-moved with knife, the hard one with gouge to level of ster-num.	None.	Death 5th day from pyemia.	Extended from an-terior surface of manubrium to left axilla; the soft tumor was in mammary glands. Hard one on manubrium.
21	König, Campe, Inaug. Diss., 1891.	M. 59	Sarcoma, right side, fist size	Resection of sternum 2d to 5th ribs, both sides; opening in right pleura 5 mark size.	None.	Died; brown atrophy of heart.	No collapse of lung (adhesions); re-spiration regular, pulse remained good; dyspnœa and feeble heart-action subse-quently.

The above table is taken, with some alterations and additions, from Keen's article in the *Medical and Surgical Reporter*, 1897, lxxvi. p. 393. The last two cases have been added. It is worthy of note that in one of Keen's cases (Case XVIII.) Cathcart's method of drainage by intermittent siphonage was employed with admirable effect in keeping the mediastinal space dry. Immediate result : Recovered, 14 ; died, 6 ; not stated, 1 ; 30 per cent. fatal. Case II. both pleura and pericardium opened, recovered, as did also Case III., with right pleura opened, but Case XXI., right pleura opened, and Case IV., both right pleura and pericardium opened ; both died. Age may have had something to do with the death of Case XXI., but the heart trouble evidently had some effect, and tuberculosis was evidently a factor in Case IV.

THIS TABLE INCLUDES ONLY

No.	Operator; year.	Bibliographic reference.	Sex and age.	Cause and duration.	Size and situation.	Rapidity of growth.	Histological classification.
1	Richerand, 1818	Paper before Royal Academy, translation by Thomas Wilson.	Adt. 40, French surgeon of Nemours.	3 years.	Large; 6th and 7th ribs, left side.	Rapid; 2 oper. before Richerand's.	Carcinoma.
2	Sedillot, 1855	Deutsch. Zeitsch. für Chir., Oct., 1898.	......			....	Sarcoma or carcinoma.
3	Schuh, 1861	Leonhard Meyer, Inaug. Diss., Erlangen, 1889.	M. 60	....	Reaching from 3d to 7th ribs, and from left sternal edge to axillary line; left side.		Chondroma.
4	Heyfelder, J. F. 1862	Beiträge zur genaueren Kenntniss der Thoraxgeschwülste. Berlin klin. Woch., 1868, v. 369.	M. age not stated.	Fracture of 7th and 8th ribs one yr. before.	13 in. in breadth, 11 in. vertically; right side.	Rapid.	Chondroma.
5	Heyfelder, J. F. 1860	Ibid.		10 years.	Filling space from vertebral column to sternum.	Slow at first, then rapid.	Chondroma.
6	Heyfelder, J. F. 1861	Ibid.	M. 33	10 years; breast inflammation	Size of fist; right side.	Slow.	Chondroma.
7	Langenbeck, 1873	Rep. by Israel, 8th German Congress for Surgery, 1879, p. 45.	F. 32	....	Size of two fists; left side.		Sarcoma.
8	Israel, 1873	Langenbeck's Arch. f. kl. Chir., 1887, Bd. xx. p.26.	F.	....	Elastic tumor of mamma size; right side.		Sarcoma.
9	Cassidy, 1874	Canada Lancet, 1874-1875. Rep. by Sylvester.	M. 29	6 years.	From size of pea grown to be size of man's fist; right side.	....	Chondroma.
10	Wattmann, 1878	Albert's Lehrbuch der Chir.	Unknown.	Unknown	Anterior chest-wall.		Chondroma.
11	Kolaczek-Fischer. 1878	Deut. Gesllsch. f. Chir., viii. 2 x, 1 80. Reported by Kolaczek	F. 48	4 years.	Left side of chest, extending from clavicle to costal arch.	....	Chondroma.
12	Tietze, 1879	Deut. Zeitschrft. f. Chir. 1891, Bd. xxxii. Rep. from Fischer's Clinic.	M. 46	Kick of hind foot of horse 3 yrs. before.	Size of child's head; right side.		Sarcoma.
13	Leisrinck, 1880	Langenbeck's Arch., Bd. xxvi. Pütt's Inaug. Diss., Berlin, 1890.	M. 37	Blow given 6 months previously.	Extending from 5th rib to lower border of thorax, and from edge of sternum to lower angle of scapula; right side.	Rapid.	Sarcoma.
14	Weinlechner 1880	Wiener med. Woch., 1882, pp. 20 and 21.	M. 37	Noted for 4 years.	Size of man's head; 3d to 5th ribs of right side.		Myxochondrosarcoma.
15	Krönlein, 1883-88	Deut. Zeitsch. f. Chir., 1893, p. 37. (Müller's article.)	F. 18	Noted 8 months.	Reached size of child's head.	Rapid.	Sarcoma.

Fig. 2.

After recovery from operation.

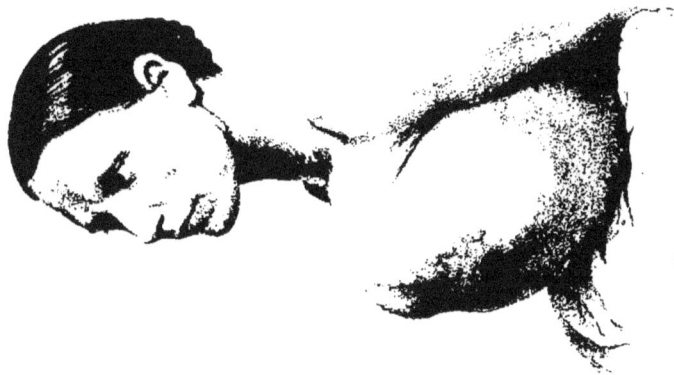

Fig. 1.

Tumor before operation.

FIG. 3.

I am indebted to Dr. Robert Fletcher, of the Surgeon-General's Library, for the reproduction of these beautiful illustrations.

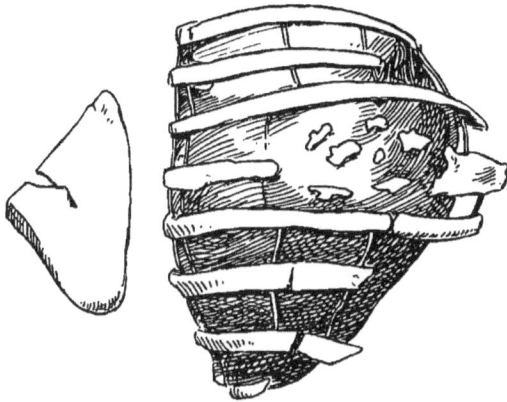

FIG. 8.

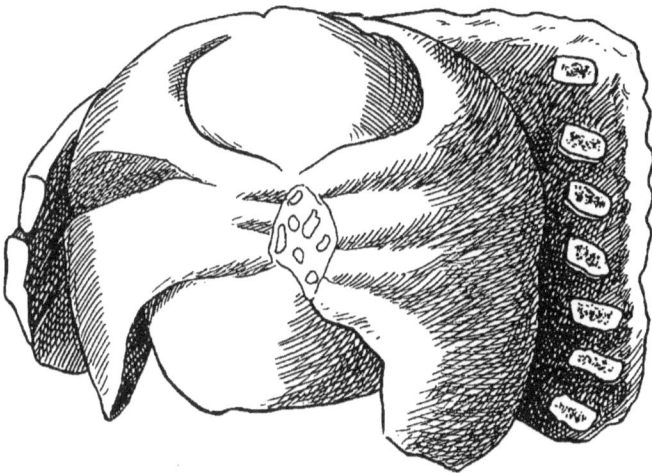

FIG. 7.

FIG. 6.

The figure represents the case after she was discharged ; unfortunately no photograph was taken before operation.

FIG. 10.

Fig. 11.

Fig. 12.

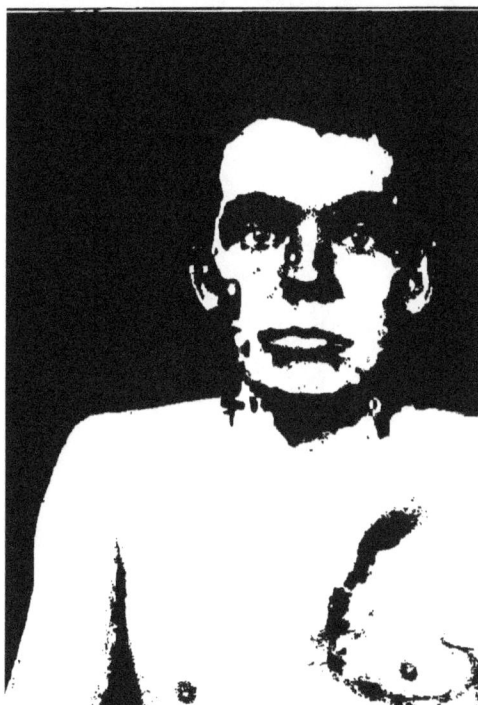

FIG. 13.

FIG. 14.

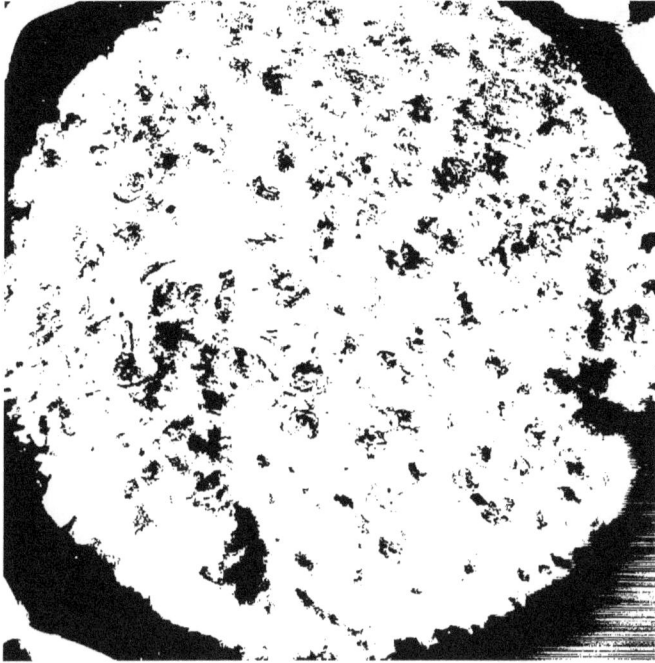

INTRAPLEURAL OPERATIONS.

Operation, anæsthetic, duration.	Size of pleural opening.	Phenomena on opening.	Management of.	Drainage and suture.	Remarks.
Crucial incision, resection of 6th and 7th ribs; no anæsthetic.	8 sq. inches.	Pneumothorax, great anxiety and oppression.	Closed with hand; oil compress.	None.	Recovery; discharged healed on 27th day. Death from recurrence 1 month later.
....	Size not stated.	....	....	....	Recovery. Remote result not stated.
....	Finger puncture.	Pneumothorax.	Obturated with finger; soft parts dropped over wound; compress.	Partial suture.	Died 2 days after operation.
7th and 8th ribs supporting tumor resected together with involved pleura.	Corresp'nding to 7th and 8th ribs.	....	....	....	Died 5 days after operation.
....	....	....	....	....	Died in first week.
Exposed 2 to 3 in. of 4th rib; pleura dissected loose.	....	....	....	....	Died on 13th day.
Involved two ribs and corresponding costal pleura.	....	....	....	....	Died from carbolic intoxication 3 months later.
....	Pleura torn in two places. Finger puncture.	....	....	....	Death followed after 3 months from exhaustion. Recovery.
....	....	....	....	....	Death from pyopneumothorax.
4th, 5th, 6th, 7th ribs and subjacent pleura.	Defect in pleura size of child's head.	Dyspnœa.	Wound sutured and drainage tube; salicylic acid solution used to wash out pleura.	....	Recovery. Recurrence one year later; death.
10 cm. of 6th, 7th, 8th, 9th ribs resected.	Pleura torn separating diaphragm from tumor	Dyspnœa slight.	....	Wound sutured and drained.	Death second day; adhesions.
6th and 7th ribs resected.	....	Collapse; pulse and respiration bad.	Ether injections; Faradization of phrenics.	Wound washed with salicylic solution and sutured with two large drains.	Died on fifth day.
3d, 4th, 5th ribs resected with involved pleura. 6th rib resected.	Pleura widely opened.	Severe collapse.	....	....	Death after 24 hours.
	....	Collapse and cyanosis.	Rapid completion of operation.	No flaps; wound packed.	Six months later piece of chest-wall, size of palm of hand, removed, together with a growth in lung the size of walnut; Mar. 1887, recurrence; 6th rib resected. August, 1898, recurrence in region of former operation. Died in 1890, of pneumonia.

No.	Operator; year.	Bibliographic reference.	Sex and age.	Cause and duration	Size and situation.	Rapidity of growth.	Histological certification.
16	Humbert, 1884	Revue de Chir., 1886, vol. vi. p. 297.	M. 21	Operated on twice in 1883.	....	....	Sarcoma.
17*	Helferich, 1885						
18	Maas, 1885	Langenbeck's Arch. f. kl. Chir., 1886, xxxiii, p. 314.	M. 42	15 years before had a blow from heavy stick	First size of hen's egg, afterward grew larger.	Slow at first, then rapid.	Chondroma
19	Schede, 1882-85	Schede's Report at Hamburg Med. Society, May 18, 1886.	F.	Operated on in 1882.	....	....	Carcinoma.
20	Trendelenburg, 1885	Leonhard Meyer, Inaug. Diss., Erlangen, 1889. Baldus, Inaug. Diss., Bonn.	M. 61	....	Tumor seated on anterior aspect of left thorax, semispherical in shape.	....	Myxochondroma.
21	Tietze, 1886-89	Deut Zeltschft. f. Chir., 1891, Bd. xxxii.	M. 66	Fall, bruising right side, in 1884.	Size of female breast.	Slowly, and with pain.	Spindlecelled sarcoma.
22	Desquin, L. 1887	Bulletin de l'Acad. Royale de Méd., Belgique, iv. s Tome ii., 1887.	M. 41	....	At first small, but after two operations grew rapidly; right side.	Slowly, and then rapidly and with pain.	Sarcoma.
23	König, 1887	Gerulanos, and Paget's Surgery of Chest, 1896.	F. 26	Rib broken 1½ years previously.	Size of fist, over 6th to 9th ribs.	....	Chondromyxosarcoma.
24	Heineke, 1889	Leonhard Meyer, Inaug. Diss., Erlangen, 1889.	M. 31	In 1887 fell strik'g side heavily.	Size of fist, on lower part of thorax.	Rapidly and painfully.	Sarcoma.
25	Müller, 1888	Paget's Surgery of Chest, p. 109.	M. 24	Noted for 4 years.	From 4th to 7th ribs, from near sternum to near axillary line.	....	Chondrosarcoma.
26	Hahn, 1888	Deutsche med. Woch., 1888, No. 50, p. 1034.	M. 33	Had been operated on twice in 1888 before.	Size of two fists.	Rapid.	Chondrosarcoma.
27	Alsberg, 1888	Riesenfeld in Deutsche med. Woch., 1889, p. 321.	F. 22	....	First noticed when of a walnut size, grew to be size of goose egg; left side.	Slow growth.	Chondroma
28	Roswell Park, 1888	Annals of Surg., 1888, vol. viii. pp. 254-57.	M. 33	Noticed a nodule 20 years ago on left leg, removed; returned, amputated; growth size of hen egg now appeared on left nipple.	....		Spindle- and round-celled sarcoma.
29	Tietze, 1889	Fischer's Surgical Clinic in Breslau. (See Case 12.)	F. 28	Blow on lt. breast at 19; a hard lump developed.	Size of man's fist. 11 cm. length. 15 cm. breadth. 12 cm. height.	Slow, then rapid.	Sarcoma.

* This case should have been included among the extra-pleural operations.

Operation, anæsthetic, duration.	Size of pleural opening.	Phenomena on opening.	Management of.	Drainage and suture.	Remarks.
7th, 8th. 9th ribs resected.	Large.	....	....	Sutured, and drain left in inferior angle	Recovery.
11 cm.of 9th, 10th, & 11th ribs with corresp. pleura resected ; chlorof.	....	Mild.	....	Wound closed with two rows of sutures.	Recovery.
Portion of thorax wall removed.	....	....	....	....	Recovery.
....	....	....	....	Drain put in and wound sutured.	Discharged cured after 3 weeks.
4th and 5th ribs resected.	....	....	....	....	Recovery ; July, 1887, small recur.; Dec. 11, recurrence, with gangrene of foot, which was removed. Feb. 1889, recur. ; death.
8th and 9th ribs with adherent pleura resected.	....	Mild.	....	....	Recovery.
8th rib resected with adherent pleura.	Freely opened.	....	....	....	Recovery. Recurrence three years later ; recovery.
7th, 8th, and 9th ribs resected with adherent pleura.	....	Mild.	Moss pillow dressing.	Sutured, and place left for drainage.	Discharged. Recur'nce July, 1888; operation; recovery.
4th, 5th, and 6th ribs resected with adherent pleura and portion of diseased lung.	Gap in pleura large as a saucer.	Stormy.	Injections of ether and camphor.	....	Recovery. Recurrence three and a half years after ; operation ; recovery.
6th, 7th, 8th, 9th, and 10th ribs, and piece of peritoneum, size of palm of hand, resected.	Pleura extensively opened.	....	....	....	Death a few days after.
8th rib resected with adherent pleura.	....	Mild.	....	Sutured, with iodoform gauze drain.	Recovery.
4th, 5th, 6th, and 7th ribs with thoracic attachments, resected. Ether.	....	Mild.	....	Sutured, and antiseptic compress.	Death.
5th rib resected with pericardial and pleural adhesions.	....	Mild.	Pleural suture.	Sutured, and iodoform dressing.	Recovery.

No.	Operator; year.	Bibliographic reference.	Sex and age.	Cause and duration.	Size and situation.	Rapidity of growth.	Histological classification.
30	Vautrin, 1889	Huitième Congrès de Chir., Lyon, 1894.	F. 19	Rt. breast enormous neoplastic mass, observed only 6 months.	It extended from sternal line to the axillary border of scapula and from clavicle to the last rib.	Rapid.	Sarcoma.
31	Bardeleben, 1890	H. Plitt, Inaug. Diss., Berlin, 1890.	M. 38	Fell on ice, struck side heavily, in 1889, 6 wks. before admission noticed several lumps in this region.	Row of tumors, the lowest on 9th rib, walnut size; the highest close to mammary line at 6th rib; bean size.	Rapid.	Sarcoma.
32	Witzel, O. 1890	Centralb. f. Chir., 1890, xvii. 523.	M. 30	....	Size of fist, situated over 10th and 11th ribs on right side.	Slow, then rapid.	Sarcoma.
33	Marsh, 1890	Brit. Med. Journ., June 14, 1890.	F. 20	For 2 years noticed lump on chest.	Growth measures 3 in. in diameter, situated over sternal ends of 3d and 4th ribs.	Slow.	Chondroma.
34	Alsberg, 1891	Caro, Deutsche med. Woch., Jan. 1893, No. 3, p. 57.	F. 31	For nearly 3 months.	Fist size; occupied space between clavicle and 4th rib and extended from right parasternal line to anterior axillary line.	Rapid.	Sarcoma.
35	Mikulicz, 1891	Reported by von Noorden, Deutsche med. Woch., 1893, xix p. 346.	M. 54	4 years.	Began size of hen's egg; at time of operation 78 cm. in circum. 35 cm. vert. meas. 35 cm. ant. post.	Slow, then rapid.	Myxochondrosarcom. cysticum.
36	König, 1891	Paget's Surgery of Chest, Case 21, p. 180.	M. 31	Only 6 wks.	Size of large apple, situated over 7th rib in axillary line.	Rapid.	Sarcoma.
37	Vautrin, 1891	Huitième Congrès de Chir., 1894, p. 162.	F. 46	Vautrin operat'd on her in Oct. 1890; recurrence 6 mos. later, came for 2d oper. Dec. 1891.	....	....	Carcinoma.
38	Zarubin, 1891	Jacobson's Oper. of Surgery, 1897, p. 581.	M. young.	7 years' standing.	21 cm. horizont. 19 cm. vertic'ly; right side of chest larger than adult head, weighed 6 lbs.	Slow.	Sarcoma.
39	König, 1893	Paget's Surgery of Chest, No. 23, p. 180.	M. 55	6 or 7 years.	4th to 7th ribs, right side.	Slow, then rapid.	Chondroma.
40	König, 1893	Ibid. p. 170.	M. 2	10 years.	Size child's head; right side in front of shoulder from 2d rib to nipple.	Slow, then rapid.	Chondroma.

Operation, anæsthetic, duration.	Size of pleural opening.	Phenomena on opening.	Management of.	Drainage and suture.	Remarks.
Chloroform. Operation lasted 1 h. 30 m. 3d, 4th. 5th, 6th, and 7th ribs resected ; wound measured 25 cm. in its transverse diameter, and vertically from a point situated 3 cm. from clavicle to 8th rib.	8 cm. by 6 cm.	Several syncopes followed.	Injections of ether.	Wound united with silkworm-gut leaving an opening 5 cm. in diameter ; flap from axilla utilized.	Recovery.
Operation lasted 90 min. 7th, 8th, and 9th ribs resected.	....	Nothing alarming.	....	A large drain put in, and wound sutured ; recovery.	
Morphine-chloroform anæsthesia.	....	Cyanosis appeared with dyspnœa at opening of pleura.	Cavity filled with warm boric solution ; drawn off.	Sutured.	Recovery. Heart suffered more than respiration.
....	3 inches of pleura.	Mild.	....	Sutured and drainage tube put in.	Recovery. Expulsion of air by forced expiration.
....	5 mark piece.	Mild.	Wo'nd stuff'd with iodoform gauze and closed by sutures, a gauze drain having been put in.	....	Discharged. Recur'nce in March, May, and June, and died Nov. 1892, of inoperable recurrence.
Morphine-chloroform, and lasted ¾ h. 7th, 8th. 9th, and 10th ribs resected.	Large, with piece of diaphragm.	No bad symptoms.	....	....	Recovery.
....	....	No serious dyspnœa.	....	....	Recovery. Rapid recurrence and death.
Chloroform. 4th, 5th, and 6th ribs resected.	3½ cm. opening in pleura.	Adhesions, only temporary embarrassment.	....	Parietal and visceral pleura united.	Recovery. Recurrence six months after operation. Woman refused further operation.
Huge mass removed, and 7th, 8th, and 9th ribs resected.	Large.	No serious disturbance.	Wound cleansed with gauze soaked in a 1 per ct. solution of boric acid.	Suture and drainage tube inserted.	Recovery.
5th and 6th ribs resected.	....	Lung collapsed.	....	Pleura sutured, hole left for drainage.	Shock ; death in a few hours.
3d rib cut thro', growth cleared with finger, piece of pleura and lung removed.	....	....	....	....	Recovery. Piece of lung also removed.

No.	Operator; year.	Bibliographic reference.	Sex and age.	Cause and duration.	Size and situation.	Rapidity of growth.	Histological classification.
41	Sheild, 1894	London Lancet, 1894, p. 741.	M. 10	12 to 15 mos.	Size of orange; right side.	Slow.	Sarcoma.
42	Faure, 1895	Revue de Chir., May 10, 1898, p. 402.	M. 61	Fract. rib, 18 months.	12 cm. diameter; right side.	Slow.	Sarcoma.
43	Karewski, 1895	Deutsche med. Woch., 1896, No. 14.	M. 36	....	3 walnut-sized in the axillary line of the right side.		Sarcoma.
44	Quénu, 1895	Revue de Chir., May 10, 1898, xv. p. 396.	M. 49	10 years.	Left side from 5th to 9th ribs.	Slow, then rapid.	Chondroma.
45	Dennis, 1896	Park's Surg., by Amer. Authors, 1896, vol. ii.	M. 15	3 months.	Right axillary aspect of chest.	Rapid.	Sarcoma.
46	Bayer, 1896	Centralblatt für Chir., No. 2, 1897.	M. 13	....	Size child's head; right thoracic wall.	....	Sarcoma.
47	Thompson, 1896	Texas Medical Journ., 1896-97, vol. xii. p. 415.	F. 45	18 months.	Size of fist; left side below and to right of nipple	Slow, then rapid.	Sarcoma.
48	Doyen, 1897	Revue de Chir., May 10, 1898, p. 399.	F. child.				Sarcoma.
49	Helferich, 1897	Gerulanos in Deut. Zeitschrift für Chir., Oct. 1898, p. 498.	M. 15	Blow, 1 yr.	Size child's head; right side.	Slow.	Sarcoma.
50	Parham, 1897	Present paper.	F. 27	7 months.	Right side, region of mamma.		Sarcoma.
51	Gauthier, 1898	Revue Internat. de Med. et de Chir., 9th year, Oct. 25, 1898, No. 20.	M. 27	4 months.	Orange size; left side posteriorly, level of 9th rib.		Sarcoma.
52	Parham, 1898	Present paper.	M. 28	6 years. Injury from brace.	Upper half of left side from clavicle to 6th rib.	Slow, then rapid.	Chondrosarcoma hyaline cartilage.

Operation, anæsthetic, duration.	Size of pleural opening.	Phenomena on opening.	Management of.	Drainage and suture.	Remarks.
Separation of scapula from thorax, unsuccessful attempt to peel tum, from pleura; accidental opening of cavity, and operation abandoned. A. C. E. mixture.	....	Moderate.	....	....	Death. Operation uncompleted.
Extirpation par morcellement. Chloroform.	2 cm.	....	Tamponed with iodoform gauze and sutured.		Recovery. Recurrence.
6th, 7th, and 8th ribs and piece of pleura resectd.	....	Great paleness and pulse imperceptible.	....	....	Recovery. Two operations.
4th, 5th, and 6th ribs resected.	Two wounds.	Cyanosis, apnœa, asphyxia, and fall of pulse.	Operation uncomplet'd; artificial respiration.	....	Collapse; death.
One rib and portion of pleura resected.	....	....	....	....	Recovery. One operation.
Pleura torn, tamponed and operation completed three days later.	Large.	Collapse.	Lung bro'ght up and fastened to edge of wound.	Opening left for drainage.	Recovery. Operation in two stages.
3d, 4th, and 5th, and cartilage of 6th rib and pleura resected; chloroform 1¼ hours.	3 inches.	Lung slowly collapsed	....	Flap sutured, iodoform gauze plug.	Gradual recovery. Recurrence five months later. Two operations
Thorax and adjacent pleura resected	....	Lung collapsed, suffocation.	Wound closed with flap.	....	Recovery.
3d, 4th, 5th, 6th, 7th, 8th, and 9th ribs resected; chloroform 1½ h.	Immense.	General collapse.	Gauze packing.	....	Death. Resection of portion of lung.
3d, 4th, and 5th resected 5 inches; chloroform 2 h.	5 inches in length.	Lung collapsed, pulse almost imperceptible.	Strychnine and digitalis hypodermatically; artificial resp.; pectoral muscle sutured over pleural rent; saline transfusion.	Gauze drainage.	Recovery. Pneumonia one year after recovery. No recurrence after twelve months.
Hand-sized portion of costal pleura resected; chloroform.	....	Breathing bad, chloroform stopped, injection of ether and caffein.	After operation injected with 1 quart artificial serum.	Sutured, with gauze pack.	Recovery. Recurrence six months after.
2d, 3d, 4th and 5th ribs resected; chloroform 1½ h.	Size of palm of hand.	Lung collapsed, pulse almost disappeared; profound shock, but quickly disappeared under forced respiration by Fell-O'Dwyer apparatus.	Lung sutured to margin of opening.	Closed without drainage; suppuration; incision.	Letter five mos. later; perfectly well. Letter May, 1899, continues well.

APPENDIX.

Gaston writes as follows in Sajous' *Annual*, 1890, Band xix.:

N. Senn, in a letter received by him December 27, 1889, reports a case of osteosarcoma of the ribs operated on by him.

He states that a man, aged about sixty-five years, noticed four months previously a small swelling on the right lower part of the chest. The growth had reached the size of a hen's egg four weeks after. It was located near the cartilage of the tenth rib, was diagnosticated as osteosarcoma, and, resection being advised, the patient entered the Milwaukee Hospital for that purpose. Ether was used for anæsthesia. On cutting down on the tumor the incisions were extended so as to remove the germ of infiltration. The rib on the vertebral side of the tumor was denuded of periosteum and divided with bone forceps about one inch (2.5 cm.) from the margins of the growth. In attempting to remove the mass with blunt instruments the pleural cavity was opened, and air rushed in, causing collapse of the lung. The tumor, with a portion of pleura, was excised, cutting through the costal cartilage with the same scissors, thus exposing the collapsed lung to view. An antiseptic compress occluded the wound while the sutures were introduced and tied. Antiseptic dressing, without drainage. The patient being threatened with collapse, the anæsthetic was suspended and stimulants administered, with a favorable result. The wound healed by primary union. The day after the operation the lung had partially expanded, and on the third day no signs of pneumothorax could be detected. The tumor, on microscopical examination, was found to be a round-celled sarcoma, intimately adherent to the costal pleura. Prior to operation adhesions should be secured by stitching the pleural surface around the tumor.

G. G. Hamilton reports, in the *Liverpool Medico-Chirurgical Journal*, July, 1892, the case of a lad, aged eighteen years, who had noticed for some time on the right side a growth about the size of a walnut, which had lately grown to the size of a turkey's egg. The growth and the eleventh rib were removed, necessitating the opening of the pleural cavity. A portion of the diaphragm had to be removed, making an opening into the peritoneal cavity. The peritoneum was stitched, the gap in the diaphragm closed, and a large drainage-tube placed in the pleural cavity, the outside wound being brought together with sutures. The wound healed by first intention, and at the end of a fortnight the lung had expanded and the lad had gone to work.

According to Sajou's *Annual*, 1889, "Special Excisions," "Cerné successfully removed the eighth rib, with the associated costal pleura, for osteosarcoma, the patient being a child." I have not been able to verify the reference, which is given as *La Normandie Médicale*, August 1, 1888.

A case is attributed also to Schnitzler, of Vienna (*Medical Press and Circular*, London, March 14, 1894; Sajous' *Annual*, 1895, Band xxviii.); but I have been unable to find this case, and the notes in the *Annual* are too inadequate for quotation.

Quénu and Longuet in their list credit two cases of fibroma of the ribs to Demarquay, stating simply that the cases were both women, and that the neighboring rib in each case was "ablated," but they give no bibliographical reference. The source of their information must have been the *Inaug. Dissertation* of Hermann Plitt, Berlin, 1890, where he says, in referring to the rarity of costal fibroma: " Demarquay saw two cases in women. These were seated in the precardial region, starting apparently from the periosteum and being connected with the pleura. Nevertheless, extirpation was done without injuring the pleura." But no details are given.

A thorough search of the Library of the Surgeon-General's office in Washington fails to disclose any case of resection of a rib for tumor to be attributed to Demarquay. All the references in the library to Demarquay are of tumors of the breast proper; in none of them was the bony wall operated on. In the *Bull. Soc. Anat. de Par.*, 1844, xviii. 331, two cases of Demarquay are related of fibro-cystic tumor of the breast in women, aged, respectively, forty-six and forty-five or fifty years, but in neither case was the bony wall of the chest disturbed. These I suspect of being the same cases as those in Quénu and Longuet's list. I have, therefore, rejected them from my list.

The following case also occurs in Quénu and Longuet's table, but a reading of it shows that there was no costal resection.

Case of Morell-Lavallée, 1861, enchondroma of the mammary region (*Bull. Soc. de Chir. de Paris*, 1862, 2 S., iii., 498-501). Enchondrome de la region mammaire, simulant une énorme tumeur du sein chez l'homme. Operation; guérison.

Male, aged thirty-five years, day laborer. No hereditary antecedents; no previous disease. Since birth the left breast had been larger than the right. Further back than he can remember the left breast had always been as large as a fist. It was about this size until about two and a half months ago, without ever having been the seat of any pain. At this time it began to grow quite rapidly, and to-day has the volume of the head of a child of four years. It has the form of a well-developed woman's breast, firm, except a prominence breaks its regularity at the antero-superior part.

He was operated on and recovered.   This case is found as
No. 10 in the list of Schläpfer von Speicher, who gives exactly the same bibliographical reference (*Gaz. des Hôp.*,
Paris, 1861, 105) as that credited by Quénu and Longuet.
This account, too, corresponds with that given in the reference quoted at the head of this abstract, and is without doubt
the same.   In both references it is distinctly stated that the
tumor was one of the soft parts.   The article in the *Gaz. des
Hôp.* gives it as a congenital tumor, cartilaginous, with colloidal change, situated under the pectoralis major and adherent
to both pectorals; origin not quite distinct.   In neither reference is it stated that any rib was resected.   We believe, therefore, that Quénu and Longuet have wrongly included this
---- in their list, and so we have excluded it from our tables.

References discovered since paper went to press:
Hartley, F., *New York Medical Journal,* April 8, 1899, p. 488.
Thoracoplasty for sarcoma of chest wall, left side.   Successful case.
No recurrence after three years and nine months.

Steele, D. A. K., *P. and S. Plexus,* Chicago, 1898-99, vol. iv. pp.
197-201.   Sarcoma of chest wall involving ninth, tenth, eleventh,
and twelfth ribs, right side, and requiring resection of a considerable
part of the diaphragm for its removal; recovery.

He was operated on and recovered. This case is found as No. 10 in the list of Schläpfer von Speicher, who gives exactly the same bibliographical reference (*Gaz. des Hôp.*, Paris, 1861, 105) as that credited by Quénu and Longuet. This account, too, corresponds with that given in the reference quoted at the head of this abstract, and is without doubt the same. In both references it is distinctly stated that the tumor was one of the soft parts. The article in the *Gaz. des Hôp.* gives it as a congenital tumor, cartilaginous, with colloidal change, situated under the pectoralis major and adherent to both pectorals; origin not quite distinct. In neither reference is it stated that any rib was resected. We believe, therefore, that Quénu and Longuet have wrongly included this case in their list, and so we have excluded it from our tables.

DISCUSSION.

Dr. WILLIS F. WESTMORELAND, of Atlanta, Ga.—This paper will revolutionize my plan of treatment in reference to surgery of the ribs. There is no question but that Dr. Parham has opened up a field which will materially change the literature of thoracotomy in regard to operations upon the ribs. From a pathological stand-point, it has not been my experience that sarcoma of the ribs follows a typical course. In operating for sarcoma in bone, unless the surgeon resects the whole bone, the operation does no good, so far as the ultimate results are concerned. I have declined to operate on such cases as the doctor has operated on. I have gotten fairly good results in some cases from simple resection of the ribs.

Another point brought out by the doctor of equal importance is the control of atmospheric influences while operating on the thoracic cavity. His results show that we can expect more from artificial respiration than we have heretofore. I have found in the majority of cases where there was no doubt about rapidly growing sarcoma on other bones, that there was no use in operating unless the patient was seen early and complete removal of the whole bone was undertaken, because extension along the medullary canal was so rapid that instead of the disease following one type, it was spindle-celled, or small, round-celled sarcoma. In making a superficial examination of the specimen that has been passed around, I do not know from what portion of the ribs the tumor was taken, but there seems to be nothing strictly indicative of the mixed variety in the whole case. As a rule, I do not think that is true of sarcoma of bone. My observations of sarcomata of bone have been that they are apt to be either of the mixed variety, or very early become so. They do not follow a typical type at all. If that is true, we can readily understand why Dr. Parham has had good results. Many of you, doubtless, know that the rib in nervous cases or in cases of insanity undergoes changes that no other bone does. It seems to me very peculiar as far as that is concerned, but that is my personal observation, and we have checked up cases from three hospitals for years. I have never found typical bone sarcomata following any particular type that had existed for any length of time. They had always been of the mixed variety. The moment the cells become typical or embryonic the disease will take on a malignancy equal to carcinoma.

The doctor's paper and specimens will change my procedure in future cases.

DR. WILLIAM SIMPSON ELKIN, of Atlanta, Ga.—I have been very much interested in Dr. Parham's paper, although I have had no personal experience in dealing surgically with sarcoma of the ribs. There was one point I was particularly interested in, namely, the pneumothorax, produced by the air entering into the pleural cavity and causing such a serious effect. Recently I operated on a case of empyema, making Estlander's operation. I removed about thirty-six inches of the ribs, from the second to the ninth inclusive. No serious trouble, whatever, was experienced in the course of the operation, until I removed the thickened pleura and intercostal muscles, when the patient came very near dying on the table. The already contracted and compressed lung became suddenly collapsed, and for a few minutes I thought my patient would die. I did not use the method of artificial respiration suggested by the essayist, but I am satisfied if I had had the apparatus at hand I would have done so and relieved myself of much anxiety. After closing the skin-flap over the abscess, and resorting to the hypodermatic use of strychnine and a normal salt solution by the rectum, the condition of my patient began to improve. For two or three days after the operation the respiration was from 40 to 60 per minute, and the pulse from 120 to 160. As soon as my patient became sufficiently conscious (about the fourth or fifth day) I had him exercise the lung by blowing through a rubber tube attached to a series of bottles. In this way the lung was gradually expanded, and the respiration soon became normal.

DR. LOUIS MCLANE TIFFANY, of Baltimore, Md.—The most excellent paper of Dr. Parham has suggested one or two points which I would like to have him answer or elaborate in his closing remarks. He spoke of removing the intercostal muscles, etc. It would be interesting to know whether in the area from which the intercostal tissues were removed the lymphatics and bloodvessels are still alive. As a rule, sarcoma generalizes itself by the veins, but sometimes it does not. In sarcoma of the long bones it is customary to take away all of the diseased portion of the bone, then there is not likely to be a recurrence.

Another point is with reference to the entrance of air into the pleural cavity. I have had decided views on this subject for some time. I am not familiar with the apparatus for artificial respiration used by Dr. Parham, but undoubtedly it is an admirable thing.

As to collapse of the lung in operations upon the chest, we have noticed a material difference between operations upon a diseased chest and a healthy one. The history which Dr. Parham gave has been repeated again and again in operations on the healthy pleura. When a patient collapses there is usually a great deal of trouble. In pleurisy the chest cavity gradually fills and the pleurisy changes to empyema, and the fluid is aspirated under anæsthesia. You are all familiar with the large trap-door operations done upon the chest, and the patient does not change respiratory action during the administration of the anæsthetic. We may do whatever we like upon a pyothorax, and there is no trouble with respiration. It is very different where the pleura is intact. This is one of the operations which should be done in two stages. Antiseptic air should be introduced into the chest for a day or two before operation, and the lung made to collapse and remain so. Dr. Murphy has been doing the same thing with nitrogen gas, and his paper has been published in the *Journal of the American Medical Association*. We know bilateral organs are luxuries; a person can get along with one ear, one leg, one ovary, and so a patient can live with one lung. I believe we can lay one-half of the chest wide open and see inside, as one looks inside the abdominal cavity, and it can be done perfectly well. In other words, produce collapse of the lung, fill the pleura with filtered air, and continue to do so. Teach the lung on that side to collapse and remain so, and then at the time of the operation it will be like an empyema.

DR. WESTMORELAND.—In how many cases have you tried this method of collapsing the lung?

DR. TIFFANY.—I have not tried it at all. The experiments of Murphy last year show that this can be done, and the future possibilities are great in this field of surgery.

DR. WILLIAM E. PARKER, of New Orleans, La.—I had the pleasure of witnessing the first of these operations by Dr. Parham. He was operating on the case at the Charity Hospital, when I happened to drop in, and I can bear testimony as to the condition of the patient. I thought he was going to die on the table. I have also had the pleasure of examining the doctor's last case, and the magnitude of the operation would be impressed upon you more deeply could you have had an opportunity of seeing the patient. I believe surgery of the chest is in its infancy, and that in the next five years great advances will be made along this line. I believe the use of the Fell-O'Dwyer apparatus will do a great deal toward advancing this line of surgery. I believe the doctor's case was the first one in which this apparatus has been used for this work.

Dr. PARHAM (closing the discussion).—Dr. Westmoreland has alluded to the mixed character of sarcomatous tumors. I found in looking over the literature of the subject and studying up the cases more thoroughly, that many of them were mixed, as chondrosarcoma, spindle-cell and round-cell sarcoma. I take this opportunity to call attention to the fact that so far as the operative diagnosis of these tumors is concerned, it is not so easy to determine what the character of the tumor is. Enchondroma is usually considered a benign tumor; it grows slowly, lasting for many years. Very strangely, many cases, as I found in looking up the literature, showed secondary growths or metastases, exhibiting many of the clinical features of malignant tumors, so that I believe enchondroma should be treated surgically as if it were a malignant tumor. Dr. Senn, in his book on tumors, says that an enchondroma very frequently takes on malignancy after operative interference not radical in character. They, therefore, should be operated on early and thoroughly. The great trouble with this whole class of tumors is this: They grow slowly, and patients do not consult us until it is too late to make a radical cure in many cases.

One of the gentlemen (Dr. Elkin) spoke of Estlander's operation; and Dr. Tiffany referred to the difference between a healthy pleura and one subjected a long time to suppuration. I agree with him that there is a great difference in this respect. Opening the normal pleura is quite a different matter from opening a pleura which has been bound down by adhesions to the lung, where the operation is much simpler.

With reference to removing the lymphatic tissue, a point referred to by Dr. Tiffany, I found that this was not done in more than a few cases. In some cases the lymphatic glands in the axilla, although not involved at the time of the operation, were subsequently removed. In one case even the supra-clavicular glands were implicated. A third operation seemed to have resulted in a cure, so far as I could learn subsequently, as there was no recurrence. In most cases the lymphatic glands are not involved. I found the statement made that in most cases the lymphatic glands were not involved and the tumor was strictly limited at its base. There was no evidence of infiltration; there were secondary tumors in but few of the cases that I could find.

With reference to taking away the whole bone, as is done in other parts of the body, it would, of course, be impracticable to do the operation if that must be done. Statistics show that by going wide of the tumor, especially in the enchrondromatous variety of tumor, a radical cure may be effected without removing the bone back to the vertebral column.

As to the Fell-O'Dwyer apparatus, my attention was again drawn to it by Dr. Matas in a paper read by him at the meeting last year of our State Medical Society, and I determined to use it if I had another case, and this case afforded the opportunity.

As to the introduction of air into the pleural cavity, which Dr. Tiffany spoke of, I have read with pleasure the paper of Dr. Murphy. It is an extremely valuable contribution to surgery in this department. The idea, of course, is not original with Murphy. Sterile air, oxygen, and even nitrogen were used before Dr. Murphy used them, and in a recent editorial in the *Medical News*, it is said that the whole apparatus for making nitrogen and introducing it into the chest cavity was invented and described long before the publication of Dr. Murphy's article. Whatever the results shall be in the treatment of tuberculosis by the injection of nitrogen into the pleural cavity, time alone will show; but so far as its usefulness is concerned as a preliminary procedure for operations upon the chest, I do not believe that it will be shown to be of much value. What we want is to get an apparatus which will prevent collapse of the lung until we can suture it to the wound. Tuffier, Quénu, and others have made valuable experiments in this respect. Tuffier has attempted to introduce a tube through the trachea or glottis above to keep the lung distended, while Quénu has attempted by a special device to hold the lung up in animals until it can be sutured. The procedure of Witzel deserves consideration, namely, filling the pleural cavity with aseptic fluid, or a boric-acid solution, and drawing it out before the last stitch is put in.

www.ingramcontent.com/pod-product-compliance
Lightning Source LLC
Chambersburg PA
CBHW021806190326
41518CB00007B/473